# CELL SIGNALLING - THERMODYNAMICS AND MOLECULAR CONTROL

Edited by **Sajal Ray**

**Cell Signalling - Thermodynamics and Molecular Control**
http://dx.doi.org/10.5772/intechopen.73411
Edited by Sajal Ray

**Contributors**

Tatsuaki Tsuruyama, Zsolt Fabian, Jan Oxholm Gordeladze, Biao Lu, Natalie Duong, Kevin Curley, Mai Do, Daniel Levy, Martha Robles-Flores, Maria Cristina Castañeda-Patlán, Gabriela Fuentes

**Notice**

Statements and opinions expressed in the chapters are these of the individual contributors and not necessarily those of the editors or publisher. No responsibility is accepted for the accuracy of information contained in the published chapters. The publisher assumes no responsibility for any damage or injury to persons or property arising out of the use of any materials, instructions, methods or ideas contained in the book.

First published in London, United Kingdom, 2019 by IntechOpen
IntechOpen is the global imprint of INTECHOPEN LIMITED, registered in England and Wales, registration number: 11086078, The Shard, 25th floor, 32 London Bridge Street
London, SE19SG – United Kingdom
Printed in Croatia

British Library Cataloguing-in-Publication Data
A catalogue record for this book is available from the British Library

Additional hard copies can be obtained from orders@intechopen.com

Cell Signalling - Thermodynamics and Molecular Control, Edited by Sajal Ray
p. cm.
Print ISBN 978-1-83880-065-9
Online ISBN 978-1-83880-066-6

We are IntechOpen,
the world's leading publisher of
Open Access books
Built by scientists, for scientists

**4,100+**
Open access books available

**116,000+**
International authors and editors

**120M+**
Downloads

Our authors are among the

**151**
Countries delivered to

**Top 1%**
most cited scientists

**12.2%**
Contributors from top 500 universities

CLARIVATE ANALYTICS
BOOK
CITATION
INDEX
INDEXED

WEB OF SCIENCE™

Selection of our books indexed in the Book Citation Index
in Web of Science™ Core Collection (BKCI)

Interested in publishing with us?
Contact book.department@intechopen.com

Numbers displayed above are based on latest data collected.
For more information visit www.intechopen.com

# Meet the editor

Sajal Ray received his MSc and MPhil degrees in Zoology and Environmental Science, respectively, from Calcutta University and was awarded his PhD from Jadavpur University. As an awardee of a Fogarty Visiting Fellowship, Dr. Ray carried out his postdoctoral research into cardiac pathology at the National Institutes of Health, Maryland, USA. His research interest is studying the immunological responses of mollusks, sponges, crabs, and earthworms exposed to common and emerging pollutants. His team is engaged in understanding the evolutionary mechanism of immunity in phylogeny. He presented his research to the World Congress of Malacology at Washington, DC. Dr. Ray, currently a professor of zoology at Calcutta University, has been teaching zoology for nearly 30 years at postgraduate level.

# Contents

# Preface

Metazoan evolution depended on intercellular coordination, which was perfected and stabilized by the functional attributes of cell signaling. Evolutionarily, the strategies and mechanisms of cell signaling bear immense significance in different tiers of physiology. This archaic process of communication is thus traced in all phyla of plants and animals. Cells are generally adapted and capable of responding to diverse forms of signals, both extrinsic and intrinsic in nature. Current trends of research indicate a variety of signaling molecules that are yet to be characterized in depth. Gases such as nitric oxide and proteins of multiple-coded identities may successfully function as signaling molecules in health, disease, and stress. Receptors of these molecules, however, may be located on the surface or interior of cells depending on the signaling strategy and mechanism. The process of cell signaling may arise from direct cell contact, cell–matrix interactions, or through secreted signaling molecules of varied molecular identities. It is assumed that successful communication among cells may be achieved through direct signaling by membrane-bound molecules or indirect signaling processes involving secreted proteins and other molecules. Intercellular signaling through cell gap junctions has been identified as another strategy of signaling, as evident in many cases. A shift in membrane-bound enzyme activity with an alteration in concentration of intracellular mediators and change in the permeability of ion channels of the cell membrane are key processes of ligand-mediated activation of cell surface receptor proteins. Secreted molecules that trigger the signaling processes follow the pathway of endocrine, paracrine, and autocrine modes as major functional strategies.

The steroid hormones thyroxin and retinoic acid are examples of signaling molecules that involve characteristic intracellular proteins functioning as receptors. Gaseous signaling molecules such as nitric oxide and carbon monoxide generally follow the paracrine mode of signaling. By influencing the activity of selected intracellular enzymes, nitric oxide alters important physiological functioning such as muscle relaxation and vascular dilation. Like signaling molecules, the pathways of intracellular signal transduction are also diverse, with multiple physiological consequences. Cell surface receptors often function as transducers by regulatory enzymes or ion channels of the cell membrane, which play an important role in communication. The mitogen-activated protein (MAP) kinase pathway of signaling plays a significant role in signal transduction and involves a family of protein kinases. These kinases are evolutionarily conserved across phyla and are indicative of their central role in metazoan evolution. In the Janus kinase/signal transducer and activator of transcription pathway, protein tyrosine phosphorylation greatly influences factor localization. Receptor tyrosine-mediated activation of the Ras-Raf-MAP kinase pathway has an immense role in the development and differentiation of multiple cell types. The embryological development of the compound eyes of *Drosophila* depends on cell signaling using this pathway.

In this edited volume, the authors highlight the importance of system biology, which is based on network theory. Most importantly, the fundamental issues of reaction dynamics and thermodynamics are stressed. A review on cell signaling is presented from the viewpoints of information thermodynamics and method quantitation. In this first chapter, encoding of signal events, code length, entropy coding, binary code theory, and adaptation of fluctuation theorem are elucidated with appropriate citations. The wnt signaling pathway, both canonical and non-canonical, have immense physiological significance. Apart from oncogenesis, its interaction with the tumor microenvironment and regulatory role in antitumor immune response are reported. In a review, the author reports on the molecular mechanism of wnt-mediated immunological cell response regulation in specific cell types. In recent years, genetically encoded reporter circuits have been revolutionizing our understanding of monitoring and manipulating different environmental stimuli. In a separate chapter, the authors describe the roles of Gaussia luciferase and a green fluorescent protein for monitoring signaling events during inflammation. They report a genetic circuit specially designed into an adeno-associated viral vector. Scientists have claimed that a combination of gluc and green fluorescent protein in a single genetic circuit as a dual reporter. It could provide an effective tool to monitor the inflammatory signals in mammalian cells. While discussing the nature of signaling during cellular metabolism, the hypoxia-related signaling process is elucidated. Oxygen is proposed as a signaling entity, which is recognized by molecular sensors conveying signals to hypoxia-inducible factors. This model is assumed to provide a better premise to understand the multilevel regulatory network operative in many cellular functional processes. In the fourth chapter, the authors mention the molecular basis of intracellular metabolic signaling in hypoxia, transcriptional feedback, and crosstalk through hypoxia-inducible factors in detail.

The signaling importance of vitamin K is discussed with reference to binding with its intranuclear receptor and subsequent activation of certain gene types. In this process, the role of steroids and xenobiotic receptors is mentioned. Growth differentiation factor 15 and stanniocalcin 2 are reported to serve as target genes of K2, which physiologically influence the γ-glutamyl carboxylase and *steroid and xenobiotic receptor*-mediated pathways in osteoblast-like cells.

**Sajal Ray**
Professor of Zoology
University of Calcutta, India

# Chapter 1

# Information Thermodynamics of Cell Signal Transduction

Tatsuaki Tsuruyama

Additional information is available at the end of the chapter

http://dx.doi.org/10.5772/intechopen.79951

### Abstract

Intracellular signal transduction is the most important research topic in cell biology, and for many years, model research by system biology based on network theory has long been in progress. This article reviews cell signaling from the viewpoint of information thermodynamics and describes a method for quantitatively describing signaling. In particular, a theoretical basis for evaluating the efficiency of intracellular signal transduction is presented in which information transmission in intracellular signal transduction is maximized by using entropy coding and the fluctuation theorem. An important conclusion is obtained: the average entropy production rate is constant through the signal cascade.

**Keywords:** information thermodynamics, fluctuation theorem, average entropy generation rate, entropy coding

## 1. Introduction

The analysis of intracellular signal transduction is one of the most important research topics in cell molecular biology. Determining the mechanisms for communicating intracellular information in the steady state, responding to changes in the external environment, and converting the change to express genetic information are a significant problem. The presented quantitative analysis may enable a comparison of signal transduction and evaluation of efficiency and should help realize the quantitative reproducibility of data for cell molecular biology and precise theoretical construction.

Gene expression cascade has been extensively studied for network study [1]. A correlation analysis of the expression pattern of a given gene is expected to give useful information for clinical diagnosis [2, 3]. Along with this evolution, protein-protein network theory has developed greatly in graph theory and phase analysis [4, 5]. Taschendorff et al. applied

IntechOpen

signaling entropy defined by correlation and transition probabilities between the proteins of interest for omics data analysis [5]. Chemokines and immunological networks are also an important theme of network research [6]. Meanwhile, considering specific reaction kinetics and thermodynamic analysis in individual reactions, there have been few studies discussing signal transduction, for example, limited to chemotaxis models of *Escherichia coli* [7], and several theoretical researches about mitogen-activated protein kinase (MAPK) cascade and bistability or ultra-sensitivity and feedback controllability of the cascade have been reported [8–13]. In addition, information thermodynamics of MAPK cascade has been recently reported [14–16]. This article reviews these recent studies from information thermodynamics in relation to fluctuation theorem.

## 2. Modeling cell signaling

### 2.1. Signaling cascade model

Intracellular signal transduction is carried out by a chain network of intracellular biochemical reactions. The network is operated by protein-protein interaction [4, 17–23]. The cell signal cascade considered here is an interesting next chain reaction mechanism: what was originally a substrate of a biochemical reaction becomes an enzyme in the next step and is a signal molecule in each step. This can be interpreted as if signal conversion is occurring rather than changing. It is possible to model this with a chemical reaction equation. The signaling step in the above cascades may be described as follows:

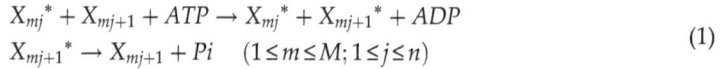

$$\begin{aligned} X_{mj}{}^* + X_{mj+1} + ATP &\rightarrow X_{mj}{}^* + X_{mj+1}{}^* + ADP \\ X_{mj+1}{}^* &\rightarrow X_{mj+1} + Pi \quad (1 \leq m \leq M; 1 \leq j \leq n) \end{aligned} \tag{1}$$

ATP, ADP, and Pi represent adenosine triphosphate, adenosine diphosphate, and inorganic phosphate, respectively. Among signal pathways, the most well-known signal pathway is the MAPK cascade. As a ligand, the epidermal growth factor (EGF) stimulates a single cell via EGF receptor (EGFR) for sequential phosphorylation of c-Raf, MAP kinase-extracellular signal-regulated kinase, and kinase-extracellular signal-regulated kinase (ERK), as shown in **Figure 1**. This cascade can transmit signal from the cell membrane to the nucleus (**Figure 1**):

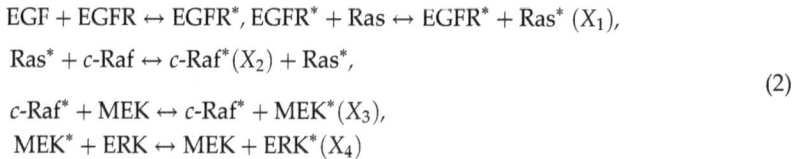

$$\begin{aligned} EGF + EGFR &\leftrightarrow EGFR^*, EGFR^* + Ras \leftrightarrow EGFR^* + Ras^* \ (X_1), \\ Ras^* + c\text{-}Raf &\leftrightarrow c\text{-}Raf^*(X_2) + Ras^*, \\ c\text{-}Raf^* + MEK &\leftrightarrow c\text{-}Raf^* + MEK^*(X_3), \\ MEK^* + ERK &\leftrightarrow MEK + ERK^*(X_4) \end{aligned} \tag{2}$$

### 2.2. Encoding of signal events

It is possible to apply information theory by considering information source coding of signal molecules. $X_j$ and $X_j^*$ represent the signal molecules. The symbol * indicates an activated state,

Extracellular stimulus

$X_1 \leftrightarrow X_1{}^*$

ATP → ADP

$X_2 \leftrightarrow X_2{}^*$

ATP → ADP

$X_3 \leftrightarrow X_3{}^*$

ATP → ADP

$X_n \leftrightarrow X_n{}^*$

DNA → mRNA

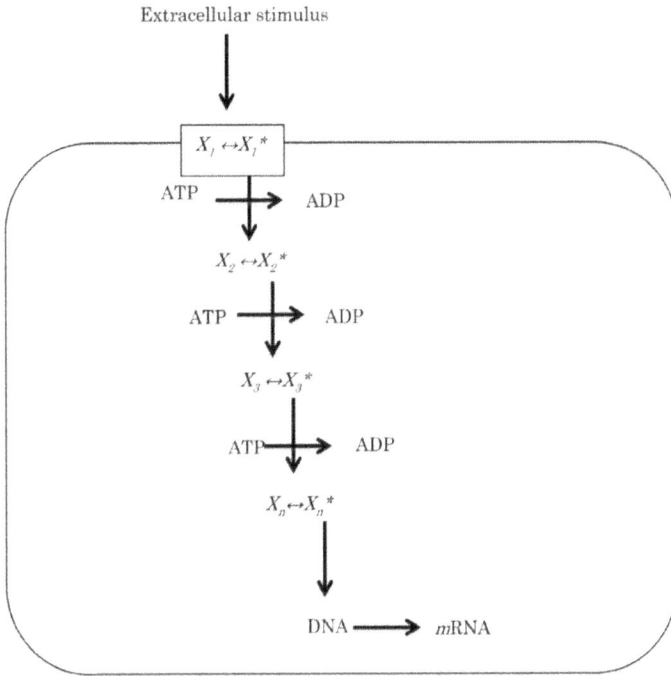

**Figure 1.** Schematic example of a signal transduction pathway. Adenosine triphosphate (ATP) represents externally supplied ATP, while the external stimulus represents the binding of a growth factor or other chemokines to a receptor. Abbreviations: ADP, adenosine diphosphate; DNA, deoxyribonucleic acid; mRNA, messenger ribonucleic acid.

and mutual conversion is possible between these two. In an actual reaction, it takes a sufficiently longer time to change from the activate state to the inactive state than it is changed from the inactive state to the active state. As shown in Eq. (1), a signal series that the activating signal molecule of step $j$-1 activates $j$ molecule is established. Following the order in which the concentration fluctuation of each signal molecule becomes significantly larger than fluctuation at the steady state,

$X_1X_2X_1{}^*X_3$... or $X_2X_3X_2{}^*X_1$... and so on.

If the probability that a signal molecule appears in one signal event is proportional to the concentration, then

$$p_j = X_j/X \tag{3}$$

$$p_j{}^* = X_j{}^*/X \tag{4}$$

with

$$\sum_{j=1}^{n} p_j + p_j{}^* = 1 \tag{5}$$

This gives $\tau$ the duration of the overall signal event, and the total number of signal events occurring during that time is taken as the total number of signal molecules X. The total signal event number $\Psi$ in a given reaction event can be described as follows:

$$\Psi = \frac{X!}{\prod_{j=1}^{n} X_j! \prod_{j=1}^{n} X_j{}^*!} \tag{6}$$

The entropy of the signal event can be defined logarithmically. The logarithm of $\Psi$ is approximated according to Starling's Equation [10]:

$$S_m = \log \psi = -X \left( \sum_{j=1}^{n} p_{mj} \log p_{mj} + \sum_{j=1}^{n} p_{mj}{}^* \log p_{mj}{}^* \right) \tag{7}$$

Here, we used Eqs. (1) and (2). This right-hand side is in the form of well-known mixed entropy. Each step of the signal pathway is considered to be a mixed state of two kinds of signal molecules.


### 2.3. Definition of code length

Here, the signal length for the time series formed by cellular signaling molecules is defined according to the theory of information source coding (**Figure 2**). $\tau_{+j}$ is the duration of the state in which the phosphorylated molecule is in an increasing state, and $\tau_{-j}$ is negative with respect to the increase in the non-phosphorylated molecule (the decline phase of the phosphorylated molecule). A positive value is assigned for $\tau_{+j}$, and a negative value is assigned for $\tau_{-j}$ giving consideration of the direction of signal transduction. For example, even if a signal is transmitted in the positive direction, if the same amount of signal is transmitted in the opposite direction, the signal becomes a net zero. To evaluate such a signal amount, the direction needs to be considered. In order to capture this, positive and negative signs are assigned to time. The definition of one total code length, i.e., total length of the given signal event, the following is given:

$$\tau_m = \sum_{j=1}^{n} \left( X_{mj} \tau_{mj} - X_{mj}{}^* \tau_{-mj} \right) \tag{8}$$

Then, (3) and (4) can be used to obtain

$$\tau_m = X \sum_{j=1}^{n} \left( p_{mj} \tau_{mj} - p_{mj}{}^* \tau_{-mj} \right) \tag{9}$$

Here, the average entropy production rate is defined during the phosphorylation or activation of signaling molecule using an arbitrary parameter $s_j$:

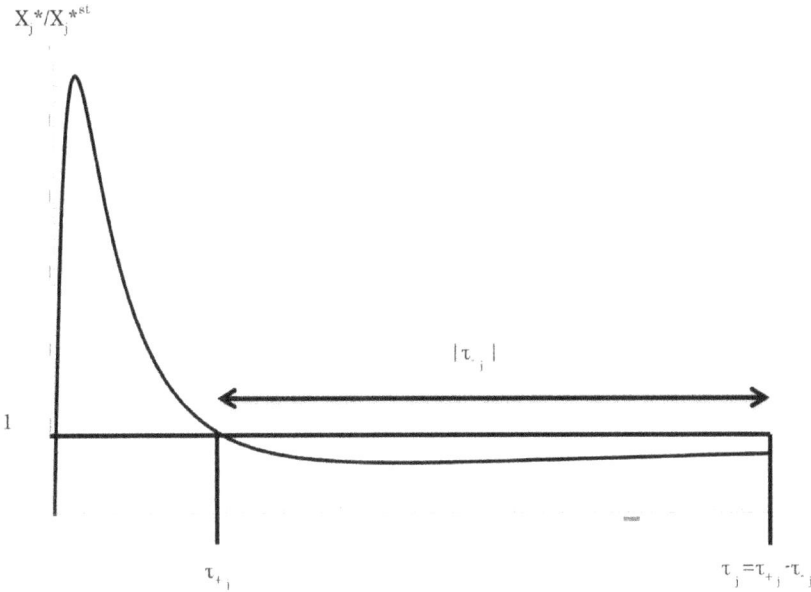

**Figure 2.** A common time course of the $j^{th}$ steps in the cascade. The vertical axis represents the ratio of the signaling active molecule concentration, $X_j^*$, to that in the steady state, $X_j^{*st}$. The horizontal axis denotes the duration (min or time unit). $\tau_j$ and $\tau_{-j}$ denote the duration of the $j^{th}$ step. Line $X_j^* = X_j^*st$ denotes the $X_j^*$ concentration at the initial steady state before the signal event.

$$\langle \zeta_j \rangle \triangleq \frac{1}{\tau_j - \tau_{-j}} \int_0^{\tau_j - \tau_{-j}} \zeta_j \left( s_j \right) ds_j \qquad (10)$$

## 2.4. Entropy coding

In order to maximize the number of signal events in a given duration, the relationship between the appearance probability (4) and the code length (9) should be calculated. The Lagrange undetermined constant method is adopted for this. If the constraint conditions are given by (5), (7), and (9), the function $G$ can be defined as follows [16]:

$$G\left(p_{m1}, p_{m2}, \cdots p_{mn}; p_{m1}{}^*, p_{m2}{}^*, \cdots p_{mn}{}^*; X\right)$$

$$= S_m - \alpha_m \sum_{j=1}^n \left(p_{mj} + p_{mj}{}^*\right) - \beta_m \tau_m \qquad (11)$$

$$= S_m - \alpha_m \sum_{j=1}^n \left(p_{mj} + p_{mj}{}^*\right) - \beta_m X \sum_{j=1}^n \left(p_{mj}\tau_{mj} - p_{mj}{}^*\tau_{-mj}\right)$$

If the partial derivative of $G$ on the right side is taken with the occurrence probability and the total number of signal molecules, respectively, then

$$\frac{\partial G}{\partial p_{mj}} = -X\left(\log p_{mj} + \beta_m \tau_{mj}\right) - \alpha_m - X \tag{12}$$

$$\frac{\partial G}{\partial p_{mj}^*} = -X\left(\log p_{mj}^* - \beta_m \tau_{-mj}\right) - \alpha_m - X \tag{13}$$

$$\frac{\partial G}{\partial X} = -\left(\sum_{j=1}^{n} p_{mj} \log p_{mj} + \sum_{j=1}^{n} p_{mj}^* \log p_{mj}^*\right) - \beta_m \left(\sum_{j=1}^{n} \tau_{mj} p_{mj} - \sum_{j=1}^{n} \tau_{-mj} p_{mj}^*\right) \tag{14}$$

If the right sides of Eqs. (12), (13), and (14) are set equal to zero, then

$$-\log p_{mj} = \beta_m \tau_{mj} \ (\tau_{jm} > 0) \tag{15}$$

$$-\log p_{mj}^* = -\beta_m \tau_{-mj} (\tau_{-mj} < 0) \tag{16}$$

and

$$\alpha_m = -X \tag{17}$$

This produces a simple result. Here, (15) and (16) are called entropy coding [16].

## 3. Information thermodynamics of cell signal transduction

### 3.1. Application of binary code theory

In practice, the signal transduction system can be classified according to two types of signaling molecules: the activated type is phosphorylated at each step of the reaction chain, and the inactive type is non-phosphorylated.

In terms of the change, the objective was to evaluate information transmission between each signal transmission step in the cascade. Increasing the active form induces the chemical potential caused by the mixed entropy change of each step in the signaling cascade and allows for biological signaling. $j$ step component is extracted from Eq. (3)

$$s_j \triangleq -k_B X\left[p_j \log p_j + p_j^* \log p_j^*\right] \tag{18}$$

Consider the entropy flow between the steps. For example, when a cell system is stimulated by the external environment or the state of a receptor at the boundary fluctuates (e.g., activation type) because of a change of the external environment, the signal cascades up to step $j$ (i.e., transmitted). In this case, because the signal is not transmitted to step $j + 1$, the concentration fluctuation of $X_j$ or $X_j^*$ differs between steps $j$ and $j + 1$, and an entropy flow can occur.

When the signal event starts and the signal is transmitted to the $j$ step, fluctuation is observed in the $j$ step as follows:

$$s_j = -k_B X\left[\left(p_j + dp_j\right)\log\left(p_j + dp_j\right) + \left(p_j^* + dp_j^*\right)\log\left(p_j^* + dp_j^*\right)\right] \qquad (19)$$

The signal has not yet reached the $j + 1$ step; hence, the entropy of the $j^{th}$ molecule remains:

$$s_{j+1} = -k_B X\left[p_j \log p_j + p_j^* \log p_j^*\right] \qquad (20)$$

We can calculate the entropy current:

$$\Delta s_j = s_j - s_{j+1} = k_B X\left(\frac{\partial s_j}{\partial p_j^*}\Delta p_j^* + \frac{\partial s_j}{\partial p_j}\Delta p_j\right) = \Delta X_j^* p_j^0 \log \frac{p_j}{p_j^*} \qquad (21)$$

Here, the logarithm of the ratio of the inactive signal molecule to the active signal molecule appeared. This form often appears. Assuming that there is no new generation of signal molecules

$$dp_j + dp_j^* = 0 \qquad (22)$$

Here, we defined the entropy current per one signal molecule, $c_j$:

$$c_j = \Delta s_j / \Delta X_j^* = p_j^0 \log \frac{p_j}{p_j^*} \qquad (23)$$

## 3.2. Fluctuation and signal transduction

Even in the steady state, signal events represented by this code sequence are occurring. When there is minor change in the extracellular environment, the amount of binding complex between the receptor on the cell membrane surface and stimulant ligand increases. This fluctuation increases the phosphorylated form of another signal molecule next to the complex and increases the fluctuation of the active type signal molecule through a chain reaction. Based on the signaling in the steady state, the increase in fluctuation indicates a signal response. In this manner, cell signaling can be distinguished as in the steady state or a fluctuation response to a change in the external environment.

## 3.3. Adaptation of fluctuation theorem to analysis of signal transduction

We defined transitional probability $p (j + 1 | j)$, which is the probability of step $(j + 1)$ given step $j$, and $p (j | j + 1)$, which is the transitional probability of step $j$ given $j + 1$ step during $\tau_j$. The logarithm of ratio $p (j + 1 | j) / p (j | j + 1)$ is divided by $\tau_j - \tau_{-j}$ and taking the limit, the AEPR from the $j$ to the $j + 1$ field satisfies the steady fluctuation theorem (FT) [24]:

$$\lim_{\tau_j - \tau_{-j} \to \infty} \frac{1}{\tau_j - \tau_{-j}} \log \frac{p(j+1|j)}{p(j|j+1)} = \zeta_j \tag{24}$$

with

$$\langle \zeta_j \rangle \triangleq \frac{1}{|\tau_j - \tau_{-j}|} \int_0^{|\tau_j - \tau_{-j}|} \zeta_j(s_j) ds_j \tag{25}$$

$$\langle \zeta_{-j} \rangle \triangleq \frac{1}{|\tau_j - \tau_{-j}|} \int_0^{|\tau_j - \tau_{-j}|} \zeta_j(s_j) ds_j \tag{26}$$

where $s_j$ is an arbitrary parameter representing the progression of a reaction event. This fluctuation theorem leads to various nonequilibrium relations among cumulates of the current. We have an equation below using signal current density [16, 24]:

$$\lim_{\tau_j - \tau_{-j} \to \infty} \frac{1}{\tau_j - \tau_{-j}} \log \frac{p(j+1|j)}{p(j|j+1)} = \frac{c_j}{k_B T (\tau_j - \tau_{-j})} \Delta X_j^* \tag{27}$$

Substituting the right side in Eq. (23) into the right side of Eq. (27), we had an important result [16]:

$$\lim_{\tau_j - \tau_{-j} \to \infty} \frac{1}{\tau_j - \tau_{-j}} \log \frac{p(j+1|j)}{p(j|j+1)} = \lim_{\tau_j - \tau_{-j} \to \infty} \frac{1}{\tau_j - \tau_{-j}} \log \frac{p_j}{p_j^*} \tag{28}$$

By substituting Eqs. (15) and (16) obtained by entropy coding on the right side of Eq. (28), using $\tau_j \ll |\tau_{-j}|$ in contrast to Eq. (27) obtains:

$$\lim_{\tau_j - \tau_{-j} \to \infty} \frac{1}{\tau_j - \tau_{-j}} \log \frac{p(j+1|j)}{p(j|j+1)} = \lim_{\tau_j - \tau_{-j} \to \infty} \beta \frac{-\tau_j - \tau_{-j}}{\tau_j - \tau_{-j}} \sim -\beta \tag{29}$$

$$\lim_{|\tau_j - \tau_{-j}| \to \infty} \frac{1}{\tau_j - \tau_{-j}} \log \frac{p(j|j+1)}{p(j+1|j)} = \lim_{\tau_j - \tau_{-j} \to \infty} \beta \frac{\tau_j + \tau_{-j}}{\tau_j - \tau_{-j}} \sim \beta \tag{30}$$

Subsequently, Eqs. (24), (29), and (30) provide

$$\beta = -\langle \zeta_j \rangle = \langle \zeta_{-j} \rangle \tag{31}$$

Accordingly, entropy coding is given using Eqs. (30) and (31):

$$-\log p_j = \langle \zeta \rangle \tau_j \tag{32}$$

$$\log p_j^* = \langle \zeta \rangle \tau_{-j} \tag{33}$$

are obtained. In conclusion, the AEPR is conserved during the whole cascade of signal transduction [16].

## 4. Conclusion

Here, each signal step is handled as actual biochemical reactions based on kinetics and thermodynamics. Signal transduction is interpreted based on encoding theory and fluctuation theorem. Regarding the relationship between information and entropy in thermodynamic mechanisms [25–27], information thermodynamics has seen remarkable developments in recent years, and information theory and thermodynamics are easier to understand when they are integrated. In particular, the theoretical results based on analysis of the Szilard engine model [28] have made it possible to compute the mutual information [29, 30] and the amount of work that can be extracted from a system by free energy changes [15]. Thus, information thermodynamics may be the theoretical basis of the signal transduction.

## Acknowledgements

This work was supported by a Grants-in-Aid from the Ministry of Education, Culture, Sports, Science, and Technology, Japan (Project No. 17013086; http://kaken.nii.ac.jp/ja/p/17013086).

## Conflicts of interest

The authors have declared no conflict of interest.

## Appendices and nomenclature

| | |
|---|---|
| MAPK | mitogen-activated protein kinase |
| EGF | epidermal growth factor |
| EGFR | EGF receptor |
| ERK | extracellular signal-regulated kinase |
| ATP | adenosine triphosphate |
| ADP | adenosine diphosphate |
| DNA | deoxyribonucleic acid |
| mRNA | messenger ribonucleic acid |
| FT | fluctuation theorem |
| AEPR | average entropy production rate |

## Author details

Tatsuaki Tsuruyama[1,2]*

*Address all correspondence to: tsuruyam@kuhp.kyoto-u.ac.jp

1 Department of Drug and Discovery Medicine, Pathology Division, Graduate School of Medicine, Kyoto University, Kyoto, Japan,

2 Clinical Research Center for Medical Equipment Development, Kyoto University Hospital, Kyoto, Japan

## References

[1] Edwards D, Wang L, Sorensen P. Network-enabled gene expression analysis. BMC Bioinformatics. 2012;**13**:167

[2] Biasco L, Ambrosi A, Pellin D, Bartholomae C, Brigida I, Roncarolo MG, Di Serio C, von Kalle C, Schmidt M, Aiuti A. Integration profile of retroviral vector in gene therapy treated patients is cell-specific according to gene expression and chromatin conformation of target cell. EMBO Molecular Medicine. 2011;**3**:89-101

[3] Dopazo J, Erten C. Graph-theoretical comparison of normal and tumor networks in identifying brca genes. BMC Systems Biology. 2017;**11**:110

[4] Teschendorff AE, Banerji CR, Severini S, Kuehn R, Sollich P. Increased signaling entropy in cancer requires the scale-free property of protein interaction networks. Scientific Reports. 2015;**5**:9646

[5] Chen CY, Ho A, Huang HY, Juan HF, Huang HC. Dissecting the human protein-protein interaction network via phylogenetic decomposition. Scientific Reports. 2014;**4**:7153

[6] Dutrieux J, Fabre-Mersseman V, Charmeteau-De Muylder B, Rancez M, Ponte R, Rozlan S, Figueiredo-Morgado S, Bernard A, Beq S, Couedel-Courteille A, et al. Modified interferon-alpha subtypes production and chemokine networks in the thymus during acute simian immunodeficiency virus infection, impact on thymopoiesis. AIDS. 2014;**28**:1101-1113

[7] Sagawa T, Kikuchi Y, Inoue Y, Takahashi H, Muraoka T, Kinbara K, Ishijima A, Fukuoka H. Single-cell *E. coli* response to an instantaneously applied chemotactic signal. Biophysical Journal. 2014;**107**:730-739

[8] Blossey R, Bodart JF, Devys A, Goudon T, Lafitte P. Signal propagation of the mapk cascade in xenopus oocytes: Role of bistability and ultrasensitivity for a mixed problem. Journal of Mathematical Biology. 2012;**64**:1-39

[9] Purutçuoğlu V, Wit E. Estimating network kinetics of the mapk/erk pathway using biochemical data. Mathematical Problems in Engineering. 2012;**2012**:1-34

[10] Zumsande M, Gross T. Bifurcations and chaos in the mapk signaling cascade. Journal of Theoretical Biology. 2010;**265**:481-491

[11] Yoon J, Deisboeck TS. Investigating differential dynamics of the mapk signaling cascade using a multi-parametric global sensitivity analysis. PLoS One. 2009;**4**:e4560

[12] Lapidus S, Han B, Wang J. Intrinsic noise, dissipation cost, and robustness of cellular networks: The underlying energy landscape of mapk signal transduction. Proceedings of the National Academy of Sciences of the United States of America. 2008;**105**:6039-6044

[13] Qiao L, Nachbar RB, Kevrekidis IG, Shvartsman SY. Bistability and oscillations in the Huang-Ferrell model of mapk signaling. PLoS Computational Biology. 2007;**3**:1819-1826

[14] Tsuruyama T. The conservation of average entropy production rate in a model of signal transduction: Information thermodynamics based on the fluctuation theorem. Entropy. 2019;**20**:e20040303

[15] Tsuruyama T. Information thermodynamics of the cell signal transduction as a szilard engine. Entropy. 2018;**20**:224

[16] Tsuruyama T. Information thermodynamics derives the entropy current of cell signal transduction as a model of a binary coding system. Entropy. 2018;**20**:145

[17] Cheong R, Rhee A, Wang CJ, Nemenman I, Levchenko A. Information transduction capacity of noisy biochemical signaling networks. Science. 2011;**334**:354-358

[18] Hintze A, Adami C. Evolution of complex modular biological networks. PLoS Computational Biology. 2008;**4**:e23

[19] Mistry D, Wise RP, Dickerson JA. Diffslc: A graph centrality method to detect essential proteins of a protein-protein interaction network. PLoS One. 2017;**12**:e0187091

[20] Imhof P. A networks approach to modeling enzymatic reactions. Methods in Enzymology. 2016;**578**:249-271

[21] Selimkhanov J, Taylor B, Yao J, Pilko A, Albeck J, Hoffmann A, Tsimring L, Wollman R. Systems biology. Accurate information transmission through dynamic biochemical signaling networks. Science. 2014;**346**:1370-1373

[22] Ito S, Sagawa T. Information thermodynamics on causal networks. Physical Review Letters. 2013;**111**:18063

[23] Barato AC, Hartich D, Seifert U. Information-theoretic versus thermodynamic entropy production in autonomous sensory networks. Physical Review. E, Statistical, Nonlinear, and Soft Matter Physics. 2013;**87**:042104

[24] Chong SH, Otsuki M, Hayakawa H. Generalized green-kubo relation and integral fluctuation theorem for driven dissipative systems without microscopic time reversibility. Physical Review. E, Statistical, Nonlinear, and Soft Matter Physics. 2010;**81**:041130

[25] Sagawa T, Ueda M. Second law of thermodynamics with discrete quantum feedback control. Physical Review Letters. 2008;**100**:080403

[26] Sagawa T, Ueda M. Generalized jarzynski equality under nonequilibrium feedback control. Physical Review Letters. 2010;**104**:090602

[27] Ito S, Sagawa T. Maxwell's demon in biochemical signal transduction with feedback loop. Nature Communications. 2015;**6**:7498

[28] Szilard L. On the decrease of entropy in a thermodynamic system by the intervention of intelligent beings. Behavioral Science. 1964;**9**:301-310

[29] Uda S, Kuroda S. Analysis of cellular signal transduction from an information theoretic approach. Seminars in Cell & Developmental Biology. 2016;**51**:24-31

[30] Uda S, Saito TH, Kudo T, Kokaji T, Tsuchiya T, Kubota H, Komori Y, Ozaki Y, Kuroda S. Robustness and compensation of information transmission of signaling pathways. Science. 2013;**341**:558-561

# Wnt Signaling as a Master Regulator of Immune Tolerance in a Tumor Microenvironment

María Cristina Castañeda-Patlán,
Gabriela Fuentes-García and Martha Robles-Flores

Additional information is available at the end of the chapter

http://dx.doi.org/10.5772/intechopen.81619

### Abstract

Aberrant Wnt signaling is a hallmark of many cancer types such as colon cancer. However, the effect of altered Wnt signaling is not only restricted to cancer cells but also dynamically interacts with a tumor microenvironment and has the ability to directly regulate the anti-tumor immune response. It has been reported that tumors induce immune tolerance through the activation of canonical Wnt signaling in dendritic cells promoting T regulatory responses, and also that both canonical and noncanonical Wnt proteins program dendritic cell responses for tolerance. Thus, the Wnt signaling pathway may be a novel and promising therapeutic target for anticancer immunotherapy. In this review, we will discuss the molecular mechanisms involved in immune cell response regulation mediated by canonical and noncanonical Wnt signaling.

**Keywords:** Wnt signaling, tumor microenvironment, immune tolerance, antitumor regulation response, regulatory T cells, dendritic cells

## 1. Introduction

Tumor-induced immune tolerance is a hallmark of cancer and constitutes a challenge for effective anticancer treatment. Overcoming immune evasion within the tumor microenvironment is crucial to successful immunotherapy to eradicate tumors.

Cancer cells operate in multiple regulatory mechanisms to evade antitumor immunity, but the signaling pathways involved in the regulation of these mechanisms are not well understood. Emerging studies have shown that Wnt signaling is a key pathway regulating tolerance versus immunity, and that it is a master regulator of T cell immune responses and of dendritic cells. In this respect, the canonical Wnt/β-catenin has been the first oncogene pathway reported that mediates immune exclusion, particularly via dendritic cells and T regulatory cells.

All of these studies suggest, therefore, that the Wnt pathway can be an attractive target to restore immune access to the tumor microenvironment.

## 2. Wnt signaling pathway

The Wnt signaling pathway is involved in the regulation of embryonic development and adult tissue homeostasis. Wnt family of secreted lipid-modified glycoproteins regulates cellular processes including stem cell maintenance, proliferation, differentiation, apoptosis, survival, cell motility, and polarity [1]. Since Wnt signals not only promote proliferation but also can control cell-fate determination and terminal differentiation in a tissue- and temporal-specific manner [2], the deregulation of Wnt signaling causes developmental defects and cancers.

In humans, 19 Wnt ligands have been identified. They bind to several receptors including Frizzled (FZD) family receptors, receptor tyrosine kinase-like orphan receptor family (ROR), low-density lipoprotein receptor-related protein co-receptors (LRP), and the related to receptor tyrosine kinase (RYK) receptor [1–3]. Wnt activates the canonical pathway that regulates transcription of target genes through the β-catenin/TCF pathway (**Figure 1**) and the noncanonical pathways that are independent of β-catenin. Disheveled (Dvl) protein is an essential element in the transduction of both canonical and noncanonical Wnt signals (**Figure 2**).

### 2.1. Canonical Wnt signaling

Canonical Wnt signals operate through regulating the phosphorylation, degradation, and localization of the transcription co-activator β-catenin (**Figure 1**). Without stimulation by Wnt, β-catenin is assembled into a destruction complex, in which APC protein plays a central role, and includes Axin, GSK-3β, and Casein kinase 1 (CK1). This complex directs a series of phosphorylation events in β-catenin mediated by CK1 and GSK-3β that targets it for ubiquitination and subsequent proteolysis via the proteasome [2, 3].

Upon binding of Wnt to FZD and LRP co-receptors, the LRP receptors are phosphorylated by CK1-alpha and GSK-3β, which recruit disheveled (DVL) and axin proteins to the plasma membrane where Dvl becomes polymerized [1–3]. The DVL polymers inactivate the destruction complex allowing β-catenin to accumulate and enter the nucleus, where it interacts with T cell factor/lymphoid enhancer factor (TCF/LEF) family members and activate a Wnt target gene program [2–4].

The tumor microenvironment (TME) contains high levels of Wnts, and aberrant β-catenin signaling occurs in many tumors. However, the effect of aberrant canonical Wnt signaling is not only restricted to cancer cells but also dynamically interacts with the microenvironment and immune system [5]. Over the last few years, it has been reported by several laboratories that Wnt signaling may also regulate T cell-mediated immune responses, and the Wnt/ β-catenin/TCF pathway in dendritic cells (DCs) plays a critical role in balancing immunity and tolerance [5].

CANONICAL WNT PATHWAY

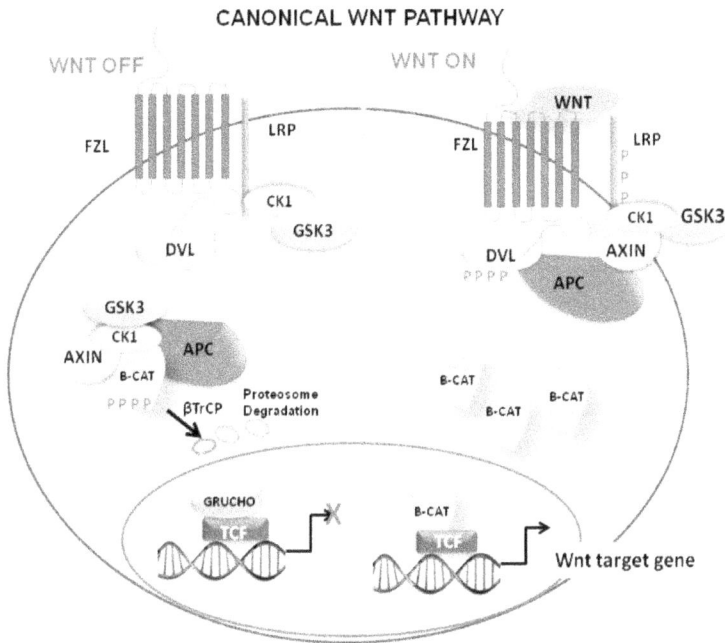

Figure 1. Canonical Wnt signaling. In the absence of Wnt ligand, β-catenin is degraded by a complex composed of Axin, APC, CK1, and GSK3. Once Wnt ligand such as archetypal Wnt3a is bound with Frizzled and LRP5/6 co-receptor, Dvl scaffolds β-catenin degradation complex resulting in accumulation of β-catenin in cytosol and nucleus. β-Catenin forms a complex with TCF to transcribe target genes.

## 2.2. Noncanonical Wnt signaling

Wnt signaling that is independent of β-catenin is referred to as the noncanonical pathway where a transcriptional response is elicited via an alternative mode of downstream signaling not involving β-catenin-TCF or β-catenin-LEF [6]. There are several noncanonical Wnt pathways, depending on the configuration of FZDs and co-receptors involved, but they can be broadly categorized into two pathways: the planar cell polarity (PCP) pathway and the Wnt/Calcium (Wnt/Ca$^{2+}$) pathway, as it can be observed in **Figure 2**.

### 2.2.1. Noncanonical PCP pathway

The PCP pathway genes code for proteins that regulate cellular polarization and directional cell movement, as initially observed during embryogenesis. The loss of normal cell polarity and adhesion, along with the acquisition of motility and invasiveness, are also fundamental steps during tumor progression and metastasis. This pathway is activated when ligands such as the noncanonical prototype ligand Wnt5a bind to FZD receptors or to FZD alternative receptors ROR1, ROR2, or RYK or through FZD with ROR or RYK as co-receptors [6, 7]. WNT/PCP signals are converted to actin cytoskeletal dynamic reorganization via the activation of small G-proteins Rac and Rho (**Figure 2**), and then, Rac and Rho activate Rho-associated kinase (Rho-kinase) and Jun N-terminal kinase (JNK)-dependent transcription [8].

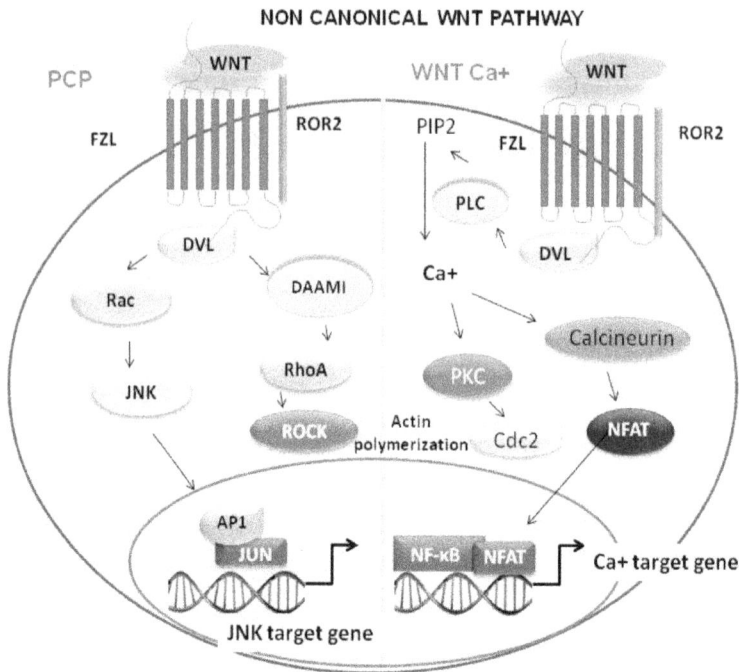

**Figure 2.** Noncanonical Wnt signaling. The activation of Frizzled by Wnt ligand such as Wnt5a is mediated by Dvl or heterotrimeric G-proteins. The two better characterized noncanonical branches are shown: planar cell polarity (at the left of the figure) in which Dvl mediates the activation of small GTPase (Rho and Rac) and JNK to promote polarized cell migration. In the other branch, the Wnt/Ca21 pathway shown at the right of the figure, archetypal noncanonical Wnt5a activates phospholipase C (PLC) to produce diacylglycerol and $Ca^{2+}$ mobilization that activates PKC isoforms, other $Ca^{2+}$-modulated kinases and calcineurin phosphatase, which in turn promotes NFAT translocation to the nucleus.

## 2.2.2. Noncanonical Wnt/$Ca^{2+}$ pathway

The Wnt/$Ca^{2+}$ pathway is initiated by the interaction of the Wnt5a/FZD receptor complex along with the participating co-receptor ROR1/2 which leads to the activation of a phospholipase C (PLC) via G protein, resulting in the production of inositol 1,4,5-triphosphate (IP3) and 1,2-diacylglycerol (DAG). IP3 induces cytosolic $Ca^{2+}$ elevation through $Ca^{2+}$ release from the endoplasmic reticulum, and both $Ca^{2+}$and DAG activate conventional and novel PKC isoforms. $Ca^{2+}$/calmodulin-dependent protein kinase II (CAMK2) and calcineurin are also representative downstream effectors of the WNT/$Ca^{2+}$ signaling cascade. In addition, WNT/$Ca^{2+}$ signaling-dependent calcineurin activation leads to dephosphorylation and subsequent nuclear translocation of NFAT for the transcriptional activation of NFAT-target genes [8].

It is well established that noncanonical Wnt ligands can antagonize the functions of canonical ligands inhibiting canonical signaling, and that there is considerable overlap between the Wnt pathway [9]. For example, CAMK2 activation leads to phosphorylation and activation of Nemo-like kinase (NLK), which can inhibit canonical Wnt/β-catenin signaling in some cells [8, 10]. However, in other contexts, Wnt5a is able to activate β-catenin-dependent transcription in the presence of Fzd4 [11]. In addition, Wnt5a can work in cooperation with the Ror2 receptor to promote β-catenin degradation independently of GSK3 and therefore inhibits Wnt3a-mediated canonical Wnt signaling [9, 12].

## 3. Wnt signaling in dendritic cell regulation

Dendritic cells (DCs) are professional antigen presenting cells acting as central players in control of innate and adaptive immunities. Emerging evidence indicates that DCs also play a pivotal role in mediating immune tolerance. Indeed, DCs are critical in maintaining tissue homeostasis by acquiring, processing, and presenting antigens to naive and resting memory T cells leading to their activation, clonal expansion, and differentiation. Paradoxically, some DCs suppress T cell responses by promoting T cell apoptosis and enhancing the development of T cells with regulatory function including CD4$^+$ Foxp3-lineage T cells (Tregs) [13].

The types of cytokines secreted by DCs dictate the outcome and type of immune response, but the receptors and signaling networks that program DCs into a tolerogenic or inflammatory state are poorly understood [14].

Tumors evade anti-tumor immunity and dendritic cells also play a key role in this. However, signaling networks for driving DC tolerogenesis in the setting of cancer remain poorly characterized. However, emerging studies has highlighted that the Wnt signaling pathway, particularly in DCs, plays a major role in regulating tolerance versus immunity. In this respect, it has been reported that Wnts in the tumor microenvironment condition dendritic cells to a regulatory state and suppress host antitumor immunity. Supporting this, Hong et al. [15] and Suryawanshi et al. [16] have shown that DCspecific deletion of Wnt coreceptors LRP5/6 in mice markedly delayed tumor growth and enhanced host antitumor immunity. These authors also found that tumors activate βcatenin/TCF4 in DCs programming them to a regulatory state, which promotes T regulatory responses while suppresses effector T cell responses [15] indicating that β-catenin activation in DCs induces regulatory T cell response and limits effector T cell response to tumors.

Spranger et al. [17] showed that a Wnt signature in cutaneous melanoma samples correlates with T cell exclusion. These authors showed that T cell priming against tumor antigens fails due to defective recruitment of CD103+ dendritic cells. In addition, β-catenin signaling downregulates the chemokine CCL4, which negatively affects the recruitment of dendritic cells to the tumor. Moreover, upregulation of IL-12 production in melanoma by increased β-catenin signaling can also lead to impaired dendritic cell maturation and induction of regulatory dendritic cells [18].

Importantly, both canonical and noncanonical (β-catenin-independent) Wnt signals have been reported to shift DCs from promoting immune responses into a tolerogenic state. In this regard, Oderup et al. [19] have found that Wnt3a activates canonical β-catenin signaling in DC, while Wnt5a triggers noncanonical signaling cascades. They found that although both canonical Wnt3a and noncanonical Wnt-5a support a tolerogenic DC phenotype, they induce distinct patterns of tolerogenic cytokine production and differential DC responses to toll-like receptors (TLRs). In addition, they showed that Wnt3a preferentially induced TGFβ, while Wnt5a was a potent stimulus for IL-10 production, and that Wnt3a but not Wnt5a, strongly stimulated DC production of VEGF-A, as well [19].

Acquired immune privilege is mediated, in part, by DCs expressing the enzyme indoleamine 2,3 dioxygenase (IDO), since IDO-expressing DCs possess potent T cell regulatory functions [13]. IDO is a heme-containing enzyme known to catalyze the rate limiting step in the degradation of the essential amino acid tryptophan to its metabolic byproducts known collectively as the kynurenines. In DCs, IDO activity has profound effects on the ability of T cells to respond to antigenic stimulation. IDO may attenuate the ability of DCs to stimulate effective T cell responses in a number of ways: T cells activated by DCs expressing IDO recognized antigen and entered the cell cycle, but IDO activity blocked subsequent cell cycle progression and enhanced T cell apoptosis [13].

Loss of TGF-β receptor III (TβRIII) expression has been shown to occur during the progression of several cancers [20]. It has been demonstrated that loss of TβRIII and the upregulation of Wnt5a by developing cancers play a role in the extrinsic control of IDO activity by local dendritic cell populations residing within tumor. These genetic changes are capable of modulating paracrine signaling pathways in the early stages of carcinogenesis to establish a site of immune privilege by promoting the differentiation and activation of local regulatory T cells [20].

It has also been reported that tumors program DCs to produce retinoic acid (RA), which promotes immune suppression by inducing T regulatory responses [21]. This is mediated through the induction of vitamin A-metabolizing enzymes via the activation of the β-catenin/TCF pathway in DCs, which in turn drives T regulatory responses and suppresses T cell effector response limiting antitumor immunity [21].

## 4. Wnt signaling in CD4 regulatory T cells

It is well known that anti-tumor T cell responses arise in cancer patients but are disabled upon tumor progression by suppressive mechanisms triggered by the interplay between malignant cells, infiltrated immune cells, and the tumor microenvironment [22].

T cells with regulatory functions (Tregs) are CD4[+] CD25[+] and express transcription factor Foxp3. They are physiologically engaged in the maintenance of immunological self-tolerance and immune homeostasis. Tumor-infiltrating Tregs can suppress effector T cells specific for tumor antigens [5].

Over the last years, it has been reported that the Wnt signal transduction pathway plays an important role both in the regulation of hematopoietic stem cell (HSC) function and in the development in the thymus, where it provides proliferation signals to immature thymocytes [22].

Accumulating experimental evidence has led to the understanding that pro-inflammatory Th17 cells are favored in their function by Wnt signaling, whereas Tregs are inhibited by canonical Wnt signaling [5]. In this respect, it has been reported that some Th17 cells are long-lived, express high levels of TCF1, and β-catenin for their self-renewal-like proliferation. Consistent with this, Gounari et al. [23] showed that Wnt-β-catenin signaling induced the expression of RORγT resulting in high amounts of IL-17 and predisposition to inflammation, colitis, and intestinal tumors. Coffer et al. [24], in contrast, showed that Wnt signaling directly modulates Foxp3 activity and thereby Treg function. TCF1 directly binds to FoxP3 and β-catenin-TCF inhibits Foxp3 transcriptional activity, thus reducing Treg-mediated suppression *in vitro* and *in vivo*. All these data therefore indicate that canonical Wnt signaling is likely a master regulatory pathway in governing the balance between Th17/Treg and thereby influences the outcome of immune responses [5].

## 5. Wnt signaling in CD8 cytotoxic T cells

The CD8+ lymphocytes act against intracellular pathogens, including viruses and bacteria or malignant cells. Upon activation, CD8+ T cells produce cytokines such as IFN-α and TNF-γ, with antitumoral or antimicrobial effects, and also produce cytotoxic perforins and granzymes, similar to Natural Killer (NK) cells. Activated CD8+ T cells also express FasL on the cell surface, which binds to its receptor, Fas, on the surface of the target cell, inducing the activation of the caspase cascade resulting in apoptosis.

A characteristic of the CD8+ T cell is the development of immunological memory, which confers protection against a secondary presentation with the same pathogen [25].

As mentioned before, Wnt signaling controls proliferation, maturation, and differentiation of T cells and dendritic cells. Staal et al. [5] have demonstrated that the Wnt-responsive transcription factors, TCF1 and LEF1, are highly expressed by naive mouse and human CD8+ T cells. The Wnt responsive transcription factor T cell factor 1 (TCF1) is well known to be critical for normal thymic T cell development. Recent studies have also revealed critical requirements for TCF1 in generation and persistence of functional memory CD8(+) T cells. Canonical Wnt signaling induced by activated β-catenin, Wnt3a canonical ligand, or GSK3β inhibitors, arrested CD8+ T cell differentiation and favored CD8+ T cell memory formation by suppressing their maturation into terminally differentiated effector T cells [26]. Importantly, constitutive activation of the canonical Wnt pathway not only favors memory CD8+ T cell formation during initial immunization but also enhanced immunity upon second encounter with the same antigen. Consistent with this, TCF1 deficiency was shown to limit the proliferation of CD8+ effector T cells and impair differentiation toward a central memory phenotype [24, 26].

Dickkopf-related protein 2 (DKK2) acts as a natural antagonist of the canonical Wnt signaling by binding to LRP5/6 co-receptor and inducing its cellular internalization. Xiao et al. [27] recently reported that the loss of adenomatosis polyposis coli (APC) in intestinal tumor cells or of the tumor suppressor PTEN in melanoma cells, upregulates the expression of DKK2, which with its receptor LRP5, provides an unusual mechanism for tumor immune evasion. DKK2 secreted by tumor cells acts on CD8+ cytotoxic lymphocytes independently of the Wnt/β-catenin pathway inhibiting STAT5 signaling by blocking STAT5 nuclear localization via LRP5. Genetic or antibody mediated inhibition of DKK2 activates natural killer (NK) cells and CD8+ T cells in tumors, inhibiting tumor progression [27].

## 6. Wnt signaling in natural killer cells

Natural killer (NK) cells are innate immune effector cells. The NK cells are derived from the hematopoietic progenitor cells (HPC) CD34+ and have been characterized by the expression of surface markers CD56+ CD3-, which have been isolated mainly from lymphoid nodules and secondary lymphoid tissues. However, a lower amount of NK cells are found in bone marrow, blood, and spleen [28] and have also been found in the skin, intestine, liver, lungs, and uterus, among other tissues [29].

It has been demonstrated the existence of a population of cells expressing NK and T cell markers, which were referred as natural killer T (NKT) cells. These cells are a relevant population of hepatic lymphocytes in both humans and mice and play important roles not only in innate defense against viral and bacterial infections but also in immune responses during carcinogenesis, autoimmunity, injury, and fibrosis [30]. Unlike T cells, NKT cells respond to lipid-based antigens: they respond to self and foreign glycolipid and phospholipid antigens presented by the MHC-I-like molecule CD1D in antigen-presenting cells (APCs), rapidly secreting the cytokines interferon gamma (IFN-γ) and IL-4 [30].

The Wnt/β-catenin signaling pathway has been implicated also in the generation of NK cells and in directing NKT cell development and functions [31]. Conditional knockout of β-catenin in mice decreases thymic NKT cell numbers, in contrast to increases in NKT cell numbers upon transgenic β-catenin overexpression [32, 33]. Consistent with this, it has been reported that the canonical Wnt inhibitor, Dickkop-1, decreased the number of NK cells in a dose-dependent manner [31].

As mentioned before, TCF/LEF mediates a nuclear response to extrinsic Wnt proteins via their binding to the co-activator β-catenin. The experimental evidence has shown that, similar to T cell maturation, TCF-1 and LEF-1 function redundantly during NK cell development: a role of LEF-1 emerges when TCF-1 levels are reduced as compared with the wild type [34]. In addition, human CD1D gene expression, which is essential for the function of NKT cells in immune regulation and surveillance of tumor cells, is regulated by LEF-1 through distal promoter regulatory elements [35].

## 7. Concluding remarks

Wnt signaling is not only restricted to cancer cells but also dynamically interacts with the microenvironment and the immune system. Dysregulation of the Wnt pathway has been implicated in many tumors, and many tumors express high levels of Wnts.

Here, we have showed that Wnt signaling is a master regulatory pathway that regulates T cell-mediated immune responses, governing the balance between activation/suppression of immune responses. As shown here, accumulating experimental evidence has demonstrated that the Wnt pathway modulates dendritic cells, CD4 T regulatory cells, cytotoxic CD8+ T cells, and NK cell functions. Thus, the Wnt signaling pathway, both canonical and noncanonical, plays pivotal roles in mediating tolerance versus immunity. Hence, blocking the Wnt pathway represents an attractive therapeutic target to overcome tumormediated immune suppression and to improve immunotherapy.

## Acknowledgements

This work was supported by grants from Universidad Nacional Autónoma de México (DGAPA-UNAM IN225717) and from CONACYT (FOSSIS 2017-289600). Gabriela Fuentes is a PhD student from Programa de Doctorado en Ciencias Biomédicas, Universidad Nacional Autónoma de México and received a fellowship from Conacyt and from DGAPA/UNAM.

## Conflict of interest

Authors declare there are no conflicts of interest.

## Author details

María Cristina Castañeda-Patlán, Gabriela Fuentes-García and Martha Robles-Flores*

*Address all correspondence to: rmartha@unam.mx

Department of Biochemistry, School of Medicine, Universidad Nacional Autónoma de México (UNAM), Mexico City, Mexico

## References

[1] Endo M, Nishita M, Fujii M, Minami Y. Insight into the role of Wnt5a-induced signaling in normal and cancer cells. International Review of Cell and Molecular Biology. 2015;**314**:117-148

[2] Zhan T, Rindtorff N, Boutros M. Wnt signaling in cancer. Oncogene. 2017;**36**:1461-1473

[3] Driehuis E, Clevers H. WNT signalling events near the cell membrane and their pharmacological targeting for the treatment of cancer. British Journal of Pharmacology. 2017;**174**(24):4547-4563. DOI: 10.1111/bph.13758

[4] Clevers H, Nusse R. Wnt/β-catenin signaling and disease. Cell. 2012;**149**:1192-1205

[5] Staal FJT, Arens R. Wnt signaling as master regulator of T-lymphocyte responses: Implications for transplant therapy. Transplantation. 2016;**100**:2584-2592

[6] Asem MS, Buechler S, et al. Wnt5a signaling in cancer. Cancers. 2016;**8**:79

[7]  Sedgwick AE, D'Souza-Schorey C. Wnt signaling in cell motility and invasion: Drawing parallels between development and cancer. Cancers. 2016;**8**:80. DOI: 10.3390/cancers8090080

[8]  Katoh M. Canonical and non-canonical WNT signaling in cancer stem cells and their niches: Cellular heterogeneity, omics reprogramming, targeted therapy and tumor plasticity (Review). International Journal of Oncology. 2017;**51**:1357-1369

[9]  Flanagan DJ, Austin CR, Vincan E, Phesse TJ. Wnt signalling in gastrointestinal epithelial stem cells. Genes. 2018;**9**:178. DOI: 10.3390/genes9040178

[10] Ishitani T, Kishida S, Hyodo-Miura J, Ueno N, Yasuda J, Waterman M, et al. The TAK1-NLK mitogen-activated protein kinase cascade functions in the Wnt-5a/Ca(2+) pathway to antagonize Wnt/β-catenin signaling. Molecular and Cellular Biology. 2003;**23**:131-139

[11] Mikels AJ, Nusse R. Purified Wnt5a protein activates or inhibits beta-catenin-TCF signaling depending on receptor context. PLoS Biology. 2006;**4**:e115

[12] Topol L, Jiang X, Choi H, Garrett-Beal L, Carolan PJ, Yang Y. Wnt-5a inhibits the canonical Wnt pathway by promoting GSK-3-independent beta-catenin degradation. The Journal of Cell Biology. 2003;**162**:899-908

[13] Huang L, Baban B, Johnson BA III, Mellor AL. Dendritic cells, indoleamine 2,3 dioxygenase and acquired immune privilege. International Reviews of Immunology. 2010;**29**:133-155

[14] Swafford D, Manicassamy S. Wnt signaling in dendritic cells: Its role in regulation of immunity and tolerance. Discovery Medicine. 2015;**19**:303-310

[15] Hong Y, Manoharan I, Suryawanshi A, Shanmugam A, Swafford D, Ahmad S, et al. Deletion of LRP5 and LRP6 in dendritic cells enhances antitumor immunity. Oncoimmunology. 2015;**5**:e1115941

[16] Suryawanshi A, Manicassamy S. Tumors induce immune tolerance through activation of b-catenin/TCF4 signaling in dendritic cells: A novel therapeutic target for cancer immunotherapy. Oncoimmunology. 2015;**4**(12):e1052932

[17] Spranger S, Bao R, Gajewski TF. Melanoma-intrinsic β-catenin signalling prevents antitumour immunity. Nature. 2015;**523**:231-235

[18] Yaguchi T, Goto Y, Kido K, Mochimaru H, Sakurai T, Tsukamoto N, et al. Immune suppression and resistance mediated by constitutive activation of Wnt/β-catenin signaling in human melanoma cells. Journal of Immunology. 2012;**189**:2110-2117

[19] Oderup C, LaJevic M, Butcher EC. Canonical and non-canonical Wnt proteins program dendritic cell responses for tolerance. Journal of Immunology. 2013;**190**:6126-6134

[20] Holtzhausen A, Zhao F, Evans KS, Hanks BA. Early carcinogenesis involves the establishment of immune privilege via intrinsic and extrinsic regulation of indoleamine 2,3-dioxygenase-1: Translational implications in cancer immunotherapy. Frontiers in Immunology. 2014;**5**:438

[21] Hong Y, Manoharan I, Suryawanshi A, Majumdar T, Angus-Hill ML, Koni PA, et al. β-catenin promotes T regulatory cell responses in tumors by inducing vitamin A metabolism in dendritic cells. Cancer Research. 2015;**75**:656-665

[22] Munn DH, Bronte V. Immune suppressive mechanisms in the tumor microenvironment. Current Opinion in Immunology. 2016;**39**:1-6

[23] Keerthivasan S, Aghajani K, Dose M, Molinero L, Khan MW, Venkateswaran V, et al. Beta-catenin promotes colitis and colon cancer through imprinting of proinflammatory properties in T cells. Science Translational Medicine. 2014;**6**:225ra28

[24] Van Loosdregt J, Fleskens V, Tiemessen MM, Mokry M, van Boxtel R, Meerding J, et al. Canonical Wnt signaling negatively modulates regulatory T cell function. Immunity. 2013;**39**:298-310

[25] Harty JT, Badovinac VP. Shaping and reshaping CD8[+] T-cell memory. Nature Reviews. Immunology. 2008;**8**:107-119

[26] Zhao DM, Yu S, Zhou X, Haring JS, Held W, Badovinac VP, et al. Constitutive activation of Wnt signaling favors generation of memory CD8 T cells. Journal of Immunology. 2010;**184**:1191-1199

[27] Xiao Q, Wu J, Wang W-J, Chen S, Zheng Y, Yu X, et al. DKK2 imparts tumor immunity evasion through β-catenin independent suppression of cytotoxic immune cell activation. Nature Medicine. 2018;**24**:262270

[28] Caligiuri MA. Human natural killer cells. Blood. 2008;**112**:461-469

[29] Carrega P, Ferlazzo G. Natural killer cell distribution and trafficking in human tissues. Frontiers in Immunology. 2012;**3**:347

[30] Gao B, Radaeva S, Park O. Liver natural killer and natural killer T cells: Immunobiology and emerging roles in liver diseases. Journal of Leukocyte Biology. 2009;**86**(3):513-528

[31] Grzywacz B, Kataria N, Niketa N, Blazar BR, Miller JS, Nandini MR. Natural killer–cell differentiation by myeloid progenitors. Blood. 2011;**117**(13):3548-3558

[32] Berga-Bolanos R, Sharma A, Steinke FC, Pyaram K, Kim YH, Sultana DA, et al. Beta-catenin is required for the differentiation of iNKT2 and iNKT17 cells that augment IL-25-dependent lung in ammation. BMC Immunology. 2015;**16**(1):62

[33] Kling JC, Jordan MA, Pitt LA, Meiners J, Thanh-Tran T, Tran LS, et al. Temporal regulation of natural killer T cell interferon gamma responses by β-catenin-dependent and independent Wnt signaling. Frontiers in Immunology. 2018;**9**:483

[34] Held W, Clevers H, Grosschedl R. Redundant functions of TCF-1 and LEF-1 during T and NK cell development, but unique role of TCF-1 for Ly49 N to K cell receptor acquisition. European Journal of Immunology. 2003;**33**(5):1393-1398

[35] Kling JC, Chen QY, Zhang T, Pincus SH, Wu S, Ricks D, et al. Human CD1D gene expression is regulated by LEF-1 through distal promoter regulatory elements. Journal of Immunology. 2010;**184**(9):5047-5054

# A Novel Genetic Circuit Supports Laboratory Automation and High Throughput Monitoring of Inflammation in Living Human Cells

Natalie Duong, Kevin Curley, Mai Anh Do, Daniel Levy and Biao Lu

Additional information is available at the end of the chapter

http://dx.doi.org/10.5772/intechopen.78568

### Abstract

Genetically encoded reporter circuits have been revolutionizing our ability to monitor, manipulate, and visualize specific cellular responses to a variety of environmental stimuli. However, the development of genetic circuits that enable both high throughput (HTP) application and laboratory automation remains challenging. In this report, we describe a novel dual-reporter circuit that utilizes a secretory Gaussia luciferase (Gluc) and a green fluorescent protein (GFP) for monitoring inflammatory signaling, a fundamental process in many life events. We designed and built this genetic circuit into a simple adeno-associated viral (AAV) vector, which is suitable for both simple transfection and efficient transduction protocols. We demonstrated high sensitivity and specificity of this new circuit and its ability to monitor a broad range of inflammatory response in various human cell models. Importantly, this novel system is simple, robust, and readily adaptable to HTP applications and laboratory automation including fluorescence activated cell sorting (FACS) and microplate reader analysis. By combining both GFP and Gluc in one genetic circuit, our new dual-reporter circuit provides an easy and powerful tool for monitoring and quantifying inflammatory signals in various mammalian cells.

**Keywords:** inflammation, NF-κB, Gaussia luciferase, TNFα, AAV, GFP

## 1. Introduction

Genetically encoded reporters such as fluorescent and bioluminescent proteins have achieved widespread success as useful research tools in life sciences, including cell biology [1–3], oncology

[4, 5], cardiology [6, 7], neurology [8, 9], as well as infection and inflammation studies [10, 11]. Because of their sensitivity in quantitative measurement, both fluorescent and bioluminescent proteins remain the top choices for monitoring live cell processes in mammals. The most commonly used reporters include green/red fluorescent proteins or their genetic derivatives (GFP, RFP, YFP or mCherry) [12–14]. Insect or marine bioluminescent proteins (Firefly luciferase or Renilla/Gaussia luciferase) are used as well [15–17]. Through codon optimization, these reporters have been genetically engineered to monitor many physiological and disease processes such as cell communications [18, 19], protein and exosome secretion [20, 21], viral infection, inflammation, and apoptosis [5, 11, 22]. In most situations, a single type of reporter may provide better sensitivity and specificity over traditional methods such as Western blot analysis or polymerase chain reactions (PCRs). For example, fluorescent proteins may provide handy real-time monitoring when a fluorescence microscope is available. However, the signal quantification is less convenient and often requires sophisticated software or expensive FACS equipment [9, 23, 24]. On the other hand, the bioluminescent reporter is easily quantifiable and can be made amenable to laboratory automation by using a less expensive luminometer [18, 25, 26]. Therefore, the combination of both types of reporters may be superior, allowing for both visual monitoring and laboratory automation.

Recently, we have developed a novel dual-reporter circuit from the marine GFP and the firefly luciferase (Fluc) and demonstrated their usage in studying gene regulation and cell signaling in mammalian cells [27]. Additionally, we and others have shown that the marine *Gaussia princeps* termed Gaussia luciferase (Gluc) has many advantages over Fluc, including secretory nature, higher assay sensitivity and specificity [17, 25, 28]. Notably, secreted Gluc activity can be easily assayed by withdrawing a portion of conditioned medium, allowing real-time monitoring while avoiding the cell lysis procedures often required for the quantification of Fluc reporter.

In this chapter, we describe the design and validation of a novel AAV vector-based dual-reporter format by a combination of GFP and Gluc for high throughput monitoring of inflammation in human cells. This new circuit has high sensitivity and specificity with little background noise in reporting inflammatory response. We demonstrate that the GFP allows for real-time monitoring and produces high-content data sets at individual cell levels using fluorescence microscope or FACS. In parallel, the secretory Gluc allows for the monitoring of inflammatory response at population levels and enables HTP analysis and laboratory automation with a luminometer or microplate reader. Together, this dual-reporter provides a robust and high throughput means to study inflammation in human cell or animal models.

## 2. Materials and methods

### 2.1. Materials

Inflammatory cytokine TNFα was obtained from R&D Systems (Minneapolis, MN). Phorbol-12-myristate 13-acetate (PMA) was obtained from MilliporeSigma (St. Lois, MO). Luciferase assay reagent and fetal bovine serum (FBS) were obtained from ThermoFisher (Waltham, MA). AAV-DJ capsid protein and the helper free viral packaging system were obtained from Cell Biolabs (San Diego, CA). The serum-free UltraCULTRE complete medium was obtained from Lonza (Anaheim, CA).

## 2.2. Genetic circuit construction

The AAV-based dual-reporter circuit was constructed by DNA synthesis and a fusion technology as previously reported [24, 29]. The reporter circuit was flanked by inverted terminal repeats (ITRs). These two ITRs were synthesized by direct DNA synthesis from Genscript (Piscataway, NJ). We then built the genetic cassette according to the configuration from the 5' to 3'-end: the transcription factor response elements (TREs) of NF-κB, a minimal CMV promoter sequences (mCMV), a dual-reporter with a self-splicing peptide (Gluc-T2A-GFP), the poly adenylation signaling sequences (Poly-A) (NF-κB reporter, GenBank Accession Number: MG786368). To evaluate the background noise, basal transcription activity and the inducible signal range, a promoterless reporter (background noise, MG786370), a minimal promoter (basal transcription activity, GenBank Accession Number: MG786371), and a full promoter CMV (signal range, GenBank Accession Number: MG786372) were built in a similar format. To assess the specificity, a dual-reporter responsive to growth factors (AP-1 reporter, GenBank Accession Number: MG786369) but not inflammatory stimuli was similarly designed and constructed. Final constructs were sequence-verified from ITR to ITR, and the annotated sequences can be retrieved from GenBank (MG786368–72).

## 2.3. Cell culture and transfection

Human embryonic kidney cells (HEK293), human liver cancer line (HepG2) and human glioblastoma line (U87) were obtained from ATCC (Manassas, VA). Cells were cultured in DMEM supplemented with 10% FBS, 2 mM GlutaMax and penicillin-streptomycin 100 U/mL. All cells were culture at 370C with 95% humid air and 5% $CO_2$.

Cell culture transfections were conducted in 6-well plates as reported [27]. Cells growing at 50–70% confluency were transfected by combining reporter DNA (1–2.5 µg/well) with Lipofectamine (Thermo-Fisher) or FuGene 6 transfection reagents (Promega) for 24–72 h. Cells were then switched to fresh medium for cytokine treatment.

## 2.4. Recombinant AAV production and titration

Reporter AAV were produced by transfecting HEK293 cells as reported [27]. Cells on culture dishes were transfected with a mixture of reporter DNA and helper AAV-reporter plasmids expressing Rep and Cap proteins. Twenty-four hours after the transfection, culture medium was changed to allow production of viral particles for additional forty-eight hours. The recombinant AAV viruses were prepared from the conditioned medium using an AAV concentration reagent according to the manufacturer's instruction (System Biosciences, Palo Alto, CA). All AAV reporter viruses were packaged with AAV-DJ capsids, which have broad tropism in transduction [30].

Viral titration and multiplicity of infection (MOI) were determined by performing green cell fluorescent assay and PCR as reported [27, 31, 32]. Briefly, HEK293 cells on 12-well plates were transduced with serial dilutions of fCMV-Gluc-T2A-GFP control viruses. After 72 h, GFP-positive cells were visually scored under a fluorescence microscope. MOI of control virus was determined by GFP positivity of the transduced cells, while the MOI of reporter viruses was estimated by comparing the relative copy numbers of the reporter viruses to those of the GFP-positive control viruses [27].

### 2.5. Gaussia luciferase assay

Gaussia luciferase activity was assayed by a luminometer (Promega, Fitchburg, WI) as previously reported [18, 24]. Briefly, conditioned medium from treated cells were collected and subsequently cleared by centrifugation at 12,000 rpm for 5 min. The cleared supernatants were used for Gluc assay. For Gluc activities quantification, 100 μL of substrate was added to 5 or 10 μL of the conditioned medium and relative light units were recorded instantaneously [25].

### 2.6. Live fluorescence microscopy

Images of living cells were typically taken using fluorescence microscopy as reported [33]. To show the intensity of GFP expression, both fluorescent and phase contrast images were recorded. To compare expression levels of GFP, identical parameters including the exposure time, contrast and gain were kept identical within each set of experiments. When the fluorescence intensity was low, images were equally adjusted to show the relative GFP intensity.

### 2.7. Fluorescence activated cell sorting analysis (FACS)

Cultured HEK293 cells were sorted and quantified by using the Accuri C6 Cytometry (BD Biosciences, San Jose, CA). More than 10,000 events were recorded via a GFP channel. Triplicate samples were analyzed to ensure consistency in results. Data were processed by CFlow Plus software.

### 2.8. GFP quantification by microplate reader

Relative GFP intensity was quantified using a Microplate Reader (BMG Labtech) following the cytokine treatment. To reduce background noise level, the conditioned media were removed and cells were washed with phosphate buffer. The relative GFP intensity was recorded for both treatment and control groups. For each sample, nine areas were measured and averaged by the OMEG 3.00R2 software.

### 2.9. Data collection and presentation

Human cells were monitored in real-time under fluorescence microscopy. Fluorescent and phase contrast images were recorded under the same experimental conditions. For Gaussia luciferase assay, GFP quantification, and FACS analysis, the data are reported as mean ± SD (n = 3), unless stated otherwise.

# 3. Results

## 3.1. Design and construction of AAV-based dual reporter circuits for monitoring inflammation in living human cells

Temporal monitoring and quantifying of inflammatory response at individual cell levels or within tissues is highly desirable [34, 35]. To accomplish this goal, we developed a new format of genetic circuit composed of transcription response elements (TREs), a minimal promoter (mCMV) and a dual-reporter (Gluc and GFP) (**Figure 1A**). According to this design, the TREs will respond to the binding of activated transcription factors such as NFκB; thus, they can switch the expression of reporter genes from an off-status to an on-status (**Figure 1A**). Since different signaling molecules may elicit distinct transcription factors (TFs). A careful choice of TREs will enable the construction of different genetic circuits for signaling monitoring. For instance, by using the binding sequences of NFκB as TREs (**Figure 1B**), this unique circuit may be able to monitor inflammation processes. Preferably, this genetic circuit will respond specifically to inflammatory molecules such as TNFα [36–38]. Taking advantage of different features of reporter proteins, the cell response can be real-time monitored by a number of HTP methods. For example, GFP may be imaged by fluorescence microscopy or quantified by microplate reader or

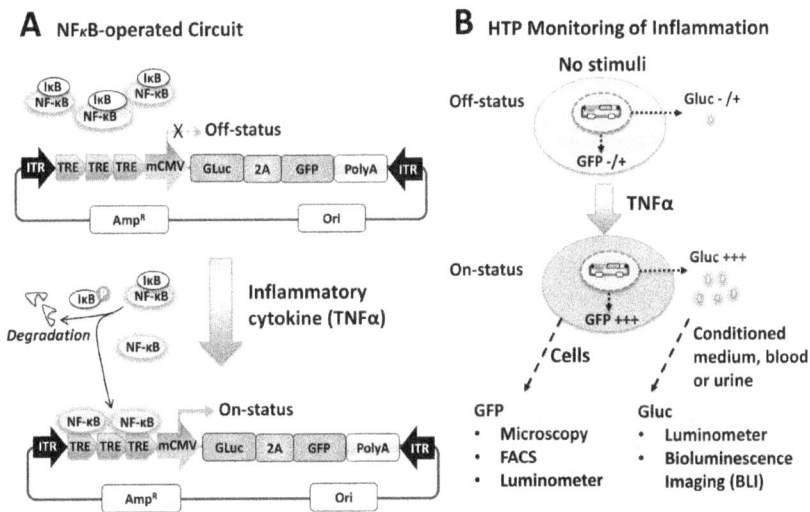

**Figure 1.** System design and workflow of AAV-based dual reporters for high throughput monitoring of inflammatory response in human cells. (A) Schematic illustration of genetic circuit of the AAV-based dual reporter. The genetic circuit (5' → 3') is composed of the transcriptional responding elements (TREs), the minimal CMV promoter (mCMV), a chimeric gene coding dual reporter proteins Gluc and GFP with a T2A (self-cleavage peptide). Depending on the availability of transcription factor (NF-κB), this inflammatory circuit may be either in an off-status with only minimal expression of reporters when no NF-κB binds to its TREs (upper panel), or in an on-status with a high level of expression when NF-κB binds to its TREs stimulated by inflammatory cytokine TNFα (lower panel). (B) Workflow for HTP monitoring of inflammatory response with GFP and Gluc. Cellular response to inflammatory stimuli can be monitored and quantified by GFP reporter (high content microscopy, FACS, microplate reader, or luminometer). Alternatively, inflammatory signaling can also be quantified by Gluc reporter from a portion of conditioned medium (Luminometer or bioluminescent imaging, BLI).

FACS. Alternatively, secreted Gluc activity can be easily quantified by assaying a portion of conditioned medium, blood, or urine. Because the wave length of Gluc emission is longer than 600 nm, Gluc becomes a preferred imager for in vivo bioluminescence imaging (BLI) (**Figure 1B**).

To test our new system, we built two reporter circuits to monitor either inflammatory processes (NFκB) or cell growth signaling (AP1). Typically, 4–6 tandem reporters of TREs can be joined together via 6-bp linkers [27]. These TREs were inserted the upstream of the mCMV-driven dual-reporter. This reporter circuit was flanked with ITR to allow AAV packaging and production.

### 3.2. Functional validation of the dual-reporter circuit

To determine the background noise, basal transcription activity, inducible signal range, and the signal-to-noise ratio of the new circuit, we further designed and constructed three additional vectors: (1) a promoterless vector to assess the background noise; (2) a minimal promoter vector to assess the basal activity; and (3) a full-length CMV promoter to assess the signal range (**Figure 2A**). We transfected these reporters into HEK293 and monitored the appearance of GFP and red fluorescent protein (RFP), which was co-transfected and served as an invariable control (driven by a constitutive EF1α promoter) (**Figure 2B**). As predicted, cells transfected by the promoterless circuit remained GFP negative for 48 h, indicating little background noise of this new AAV reporter circuit (**Figure 2B, a, g**). For the minimal promoter circuit, few cells were weakly positive for GFP, indicating low levels of basal expression (**Figure 2B, b, h**). In contrast, ~80% of cells exhibited strong GFP fluorescence in the fCMV group (24–48 h) (**Figure 2B, c, i**). However, under the same experimental condition, the steady expression levels of control RFP remained consistent among different groups for both 24 h (**Figure 2B, d–f**) and 48 h (**Figure 2B, j–l**), suggesting the differential expression of GFP was attributed to the promoter usage rather than the differences caused by transfection discrepancy. Together, these data validated the functionality of our new circuit and confirmed that GFP could be used for reporting signaling strength in living human cells.

In parallel, we also examined the Gluc activities from the conditioned media to determine if the Gluc activities were similarly regulated depending on the promoter types. As shown in **Figure 2C**, the promoterless circuit showed a very low background noise while the mCMV promoter circuit exhibited a significant increase in Gluc activities (~30.9-fold at 24 h, ~56.9-fold at 48 h). A marked ~2255- or ~3847-fold increase in Gluc activity was recorded for the fCMV promoter circuit. These data confirmed that the new reporter format has low levels of background noise and a broad signal range with a signal-to-noise ratio of ~3847:1.

### 3.3. Specificity of the dual-reporter circuit in monitoring inflammation using transfection protocol

Next, we tested whether the dual-reporter system was specific in monitoring inflammation with a commonly used transfection protocol. We conducted a comparative study on two distinctive pathways, the inflammation (NFκB) and cell growth (AP-1). These two pathways have been

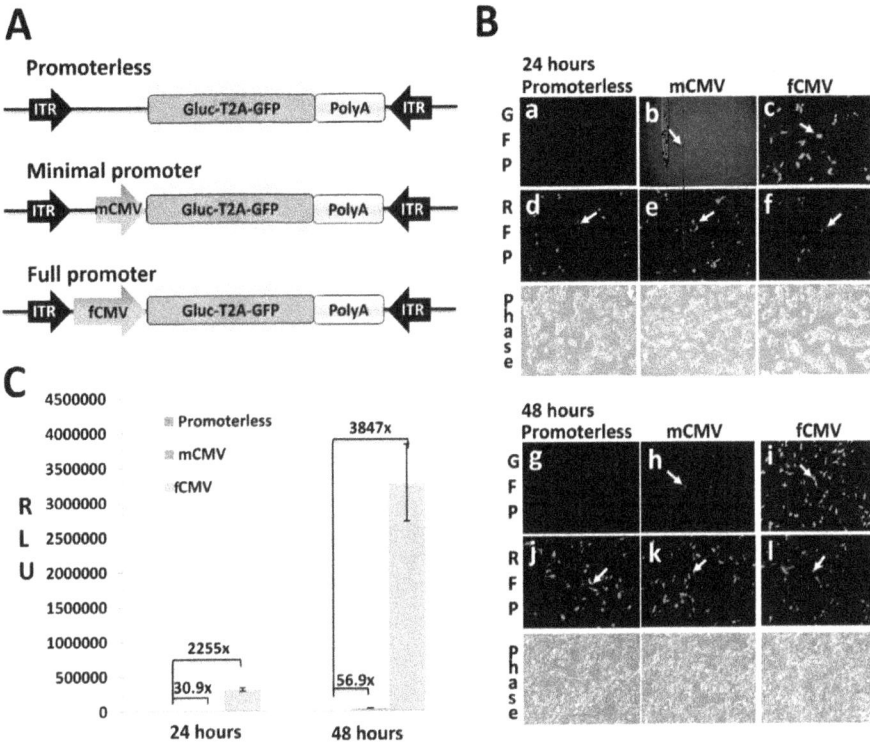

Figure 2. System setup and performance analysis. (A) Design and construction of three AAV-based dual reporters for system testing. The dual-reporter circuit is flanked by inverted terminal repeats (ITR), which allows for packaging into recombinant AAV. Promoterless, minimal promoter (mCMV) and full CMV (CMV)-driven dual reporter circuit are shown from top to bottom. (B) Expression of dual-reporters in live HEK293 cells. The GFP expression (green) in HEK293 cells were recorded with a fluorescence microscope following transfections of either a promoterless, an mCMV, or an fCMV-driven reporter at 24 h (a–c) and 48 h (g–i), which were co-transfected with a positive control plasmid DNA expressing RFP (d–f for 24 h; j–l for 48 h). Arrows indicate GFP- or RFP-positive cells. The bottom panels show the corresponding phase-contrast images for each group. (C) The Gluc activity from the conditioned medium was determined by a luciferase assay following the transfection of three reporters at the same time points. The luciferase activity was expressed as relative light units (RLU), normalized against protein input, and presented as fold increase over untreated control (mean ± SD, n = 3) with statistical significance of P<0.001, using student's T-test. The bottoms panels are representative phase-contrast images for each treatment group.

shown to be specifically activated by proinflammatory cytokine (TNFα) and cancer promoting reagent (PMA), respectively [18, 39]. Accordingly, we co-transfected HEK293 cells using each of these reporters along with EF1α-driven RFP as invariable reference to determine specific effects of TNFα and PMA on reporter activations. After cells were transfected with NFκB-Gluc-2A-GFP reporter, treatment of cells with 10 ng/mL TNFα induced a marked increase in GFP levels (Figure 3A, b). In parallel, a 42-fold increase in Gluc activity was detected, indicating an activation of the inflammatory pathway (Figure 3B). As expected, very few GFP-positive cells were present in the wells treated with PMA, suggesting specific activation of NFκB by TNFα but not PMA (Figure 3A, c). Again, no significant change in Gluc activities was observed in mock

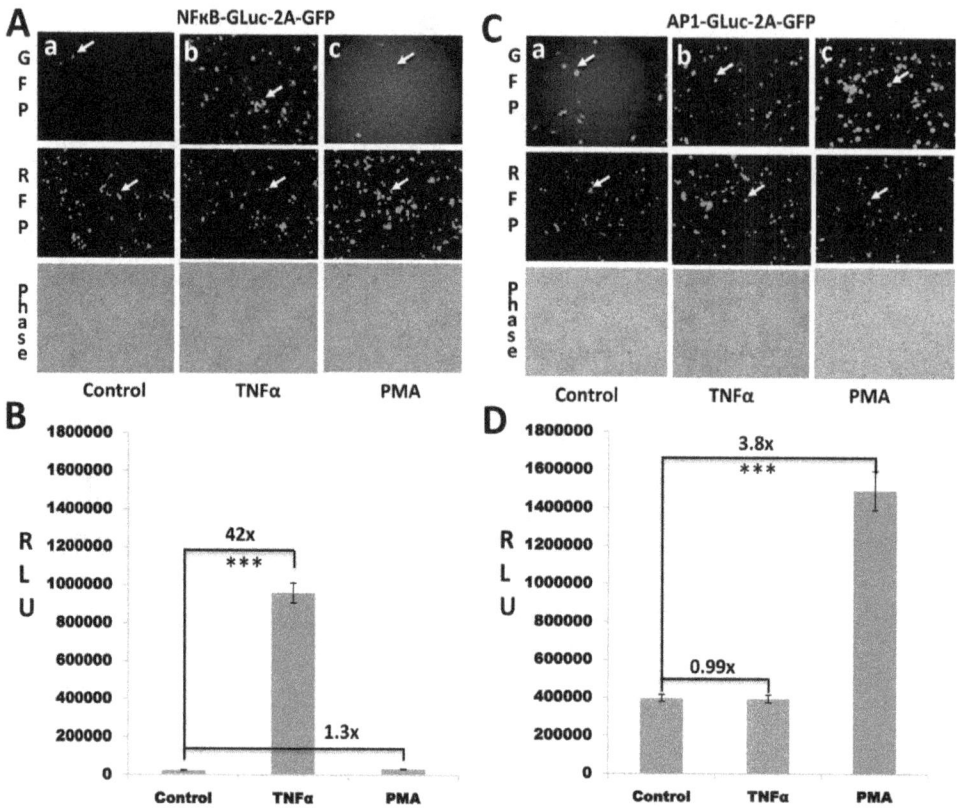

Figure 3. The specificity of the AAV-based dual-reporters by transfection. HEK293 cells were transfected with NF-κB reporter for 24 h. Cells were then switched to low serum medium in the presence or absence of either TNFα (10 ng/mL) or PMA (50 ng/mL) for 24 h. The images of GFP (A: a–c), RFP (A: d–e), and phase (A: lower panel) were recorded 24 h after the treatment. The corresponding luciferase activity was assayed using the conditioned medium collected from a control, TNFα or PMA treatment group (B). In a separate set of experiments, HEK293 cells were transfected with AP-1 reporters for 24 h, followed by the treatment of either TNFα (10 ng/mL) or PMA (50 ng/mL) for additional 24 h. The GFP (B: a-c), RFP (B: d-e), and phase (B: lower panel) images were recorded and the corresponding luciferase activity was also assayed and graphed (D). Arrows indicates the GFP- or RFP-positive cells. The luciferase activity was expressed as relative light units (RLU) and presented as fold increase over untreated control (mean ± SD, n = 3). *** denotes P<0.001, using student's T-test.

control or cells treated with PMA (**Figure 3B**). Conversely, after cells were transfected with the AP1-Gluc-2A-GFP reporter, treatment of cells with 50 ng/mL PMA induced a significant increase in GFP levels (**Figure 3C, c**), in line with a 3.8-fold increase in Gluc activities (**Figure 3D**). In contrast, neither mock-control (**Figure 3C, a**) nor TNFα (**Figure 3C, b**) caused any significant changes in the expression levels of GFP or Gluc (**Figure 3D**). Additionally, the co-transfected invariable reference RFP showed little changes among different groups (**Figure 3A, C**, middle panels), excluding the possibility that such differences were caused by discrepancies in transfection efficiency. Taken together, these results confirm the specificity of our new reporter circuit in monitoring inflammation using a simple transfection protocol.

## 3.4. Monitoring cell signaling response using an AAV transduction protocol

Following successful monitoring of cell signaling with a transfection protocol, we further test whether the new AAV-based circuit could be successfully packaged into delivery viral particles to deliver the reporter circuit to various cells [40–42]. For virus packaging, we used our established protocol to generate recombinant reporter viruses [27]. For cell transduction, packaged viruses (∼ 1× MOI viruses) were used to transduce HEK293 cells for 24 h. Following the transduction, cells were washed and switched to UltraCULTURE for the treatment of TNFα or PMA. Similar to the transfection experiment, the NFκB reporter circuit showed a marked increase in Gluc activity (85-fold over control) in response to the TNFα treatment but not PMA (**Figure 4A**). Conversely, the AP-1 reporter circuit showed a smaller but significant increase in Gluc activity (9.6-fold over control) for the treatment of PMA but not TNFα (**Figure 4B**). These results validated the specificity of this dual-reporter using a highly effective

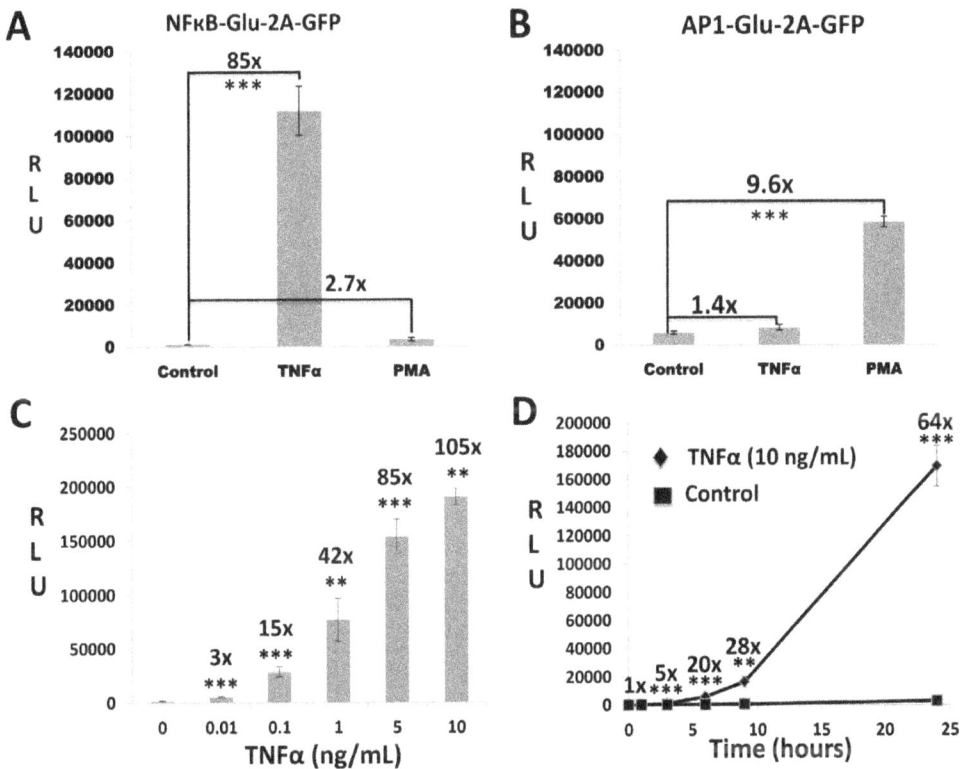

**Figure 4.** The specificity of AAV-based reporters by transduction. HEK293 cells were transduced with either NFκB (A) or AP-1 (B) reporter AAV (MOI = 1) for 24 h. Cells were then switched to low serum medium in the presence or absence of TNFα (10 ng/mL) or PMA (50 ng/mL) for additional 24 h. In separate experiments, following 24 h transduction with NFκB reporter AAV, HEK293 cells were subject to either the treatment of increasing concentration of TNFα (0, 0.1, 1, 5 10, and 50 ng/mL) for 24 h (C), or in the presence of TNFα (10 ng/mL) for 0, 3, 7, 24, 48, and 72 h (D).The luciferase activity was determined using conditioned medium, and expressed as relative light units (RLU) or presented as fold increase over untreated control (mean ± SD, n = 3). ** denotes $P<0.01$, while *** denotes $P<0.001$, using student's T-test.

transduction protocol. Consistent with low basal level of Gluc activities, we also observed a low GFP expression in control cells, which increased in their intensities in response to TNFα treatment but not to PMA. Likewise, the GFP intensity increased in response to PMA but not to TNFα treatment in the AP-1 system, indicating specific response to its corresponding singling stimuli. It is worth noting that both GFP intensity and Gluc activity appeared to be low for cells with transduction in comparison to transfection. However, the relative fold-increase of Gluc appeared to be more prominent (85-fold vs. 42-fold increase for NF-κB and 9.6-fold vs. 3.8-fold increase for AP-1), which suggests that the lower background noise may increase the detection sensitivity using viral transduction protocol. Nevertheless, our data support the notion the new format of reporter circuit could be used to monitor and quantify inflammation or cell growth signaling by either simple transfection or highly efficient transduction protocols.

### 3.5. Sensitive and real-time monitoring of inflammatory response

We next performed the dose-response and time-course experiments to examine the sensitivity of our inflammatory reporter circuits in HEK293 cells. Following the transduction of cells with either AAV-based NFκB reporter or control plasmid for 24 h, cells were treated with incremental amount of TNFα (0.01, 0.1, 1, 5, and 10 ng/mL) for 24 h. As low as 0.01 ng/mL TNFα induces a significant rise of Gluc activity (3-fold over control, **Figure 4C**). As expected, higher TNFα concentrations (0.1–10 ng/mL) elicited more robust responses (15–105-fold over control) in a dose-dependent fashion (**Figure 4C**). In agreement with Gluc activities, GFP images exhibited a similar pattern of dose-dependent increase of GFP expression levels in responding to TNFα.

To temporally examine the activation of the NFκB reporter, we conducted a time-course study on the effects of TNFα using the established transduction protocol. We used a dosage of 10 ng/mL of TNFα because this dosage can induce a robust response in the above dose-response experiments. As early as 3 h following TNFα treatment, Gluc activity started to rise significantly (5-fold over control, **Figure 4D**). During the first 24 h, Gluc activity steadily increased (5–64-fold over control) (**Figure 4D**). Together, these results show that our new dual-reporter responds to inflammatory stimuli in both a dose- and time-dependent fashion, hence demonstrated the usefulness of the secreted Gluc for temporal monitoring of inflammation with our new reporter circuit in living human cells.

### 3.6. Coupling of dual-reporter with cytometer and automation-compatible microplate reader

To further explore whether our new dual-reporter circuit is amenable to HTP applications, we conducted a series of high content experiments that consisted of multiple dose- and time-course studies. Following the transduction of AAV-based NFκB reporters, cells were treated with incremental concentrations of TNFα (0, 0.01, 0.1, 1, 5, 25, 100 ng/mL). At multiple time-points (0, 1, 3, 6, 9, 24, 32, and 48 h), a small portion of culture medium was collected to determine the Gluc activity. As shown in **Figure 5A**, both time- and dose-dependent responses to the treatment of TNFα were detected. However, the cellular response to TNFα appeared to be attenuated in the following 24–48 h, indicating the maximum possible stimulation achieved at 24 h time-point. It is important to note that cellular response demonstrated a typical early response (**Figure 5B**). At doses of 0.1 ng/mL and above, significant increases (**Figure 5B**) in Gluc activities

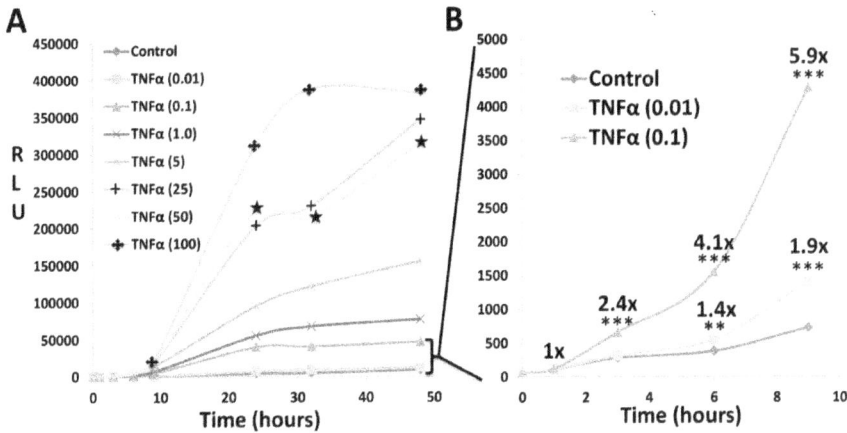

**Figure 5.** (A) HEK293 cells respond to TNFα stimuli in a time- and dose-dependent manner. Following transduction with NF-κB reporter virus (MOI = 1) for 24 h, HEK293 cells were treated with increasing concentration of TNFα (0, 0.01, 0.1, 1, 5, 10, 25, 50 and 100 ng/mL). A small portion of conditioned medium was collected at the indicated time-points (0, 1, 3, 6 and 9 h). The luciferase activity was determined using the conditioned medium, and expressed as relative light units (RLU) or presented as fold increase over untreated control (mean ± SD, n = 3). (B) Inset of control and treatment concentrations of 0.1 and 0.01 ng/ml. ** denotes P<0.01, while *** denotes P<0.001, using student's T-test.

were apparent. Even at the lowest dose of 0.01 ng/mL, a small increase (1.4-fold over control) in Gluc activities was detected at 6 h (**Figure 5B**), supporting a receptor-mediated quick activation model. Additionally, the secreted reporter Gluc enabled us to conduct these multi-dose (eight dosages) and multi-times (8 time-points) experiments in a triplicate format (three biological repeats), yielding a total of 192 data points, rendering a reliable and informative pattern of response.

We next examined the individual cell response to inflammatory cytokines using FACS as a HTP tool. HEK293 cells were transduced with AAV-based NFκB reporters, and the GFP emission was quantified by FACS. As expected, cells in the background group (mock transfection) were GFP-negative (**Figure 6A**, left panel), while reporter-transduced cells showed weak GFP expression in the absence of TNFα (**Figure 6A**, middle panel), but high GFP expression in the TNFα-treatment group (**Figure 6A**, right panel), demonstrating a robust cellular response. In parallel, our cytometry data revealed high intensity of GFP signal at individual cell levels: 0% in background group vs. 26% in control group vs. 83.6% in TNFα group (**Figure 6B**). A marked shift of GFP intensity following TNFα treatment was apparent when these graphs were merged (**Figure 6C**). Additionally, the individual response could also be summarized and averaged to evaluate cellular response as a heterogeneous population, which showed ~40-fold increase in TNFα group over control (**Figure 6D**). Together, our results demonstrated that flow cytometry can be used to assess inflammatory response at both individual cell and population levels using our novel reporter circuit.

Next, we examined whether a more commonly available microplate reader can be an alternative readout tool compared to the expensive cytometer for GFP quantification. Following the transfection of HEK293 cells with our NFκB reporter, the GFP intensity was recorded by fluorescent microscopy and the GFP emission was quantified with a microplate reader equipped

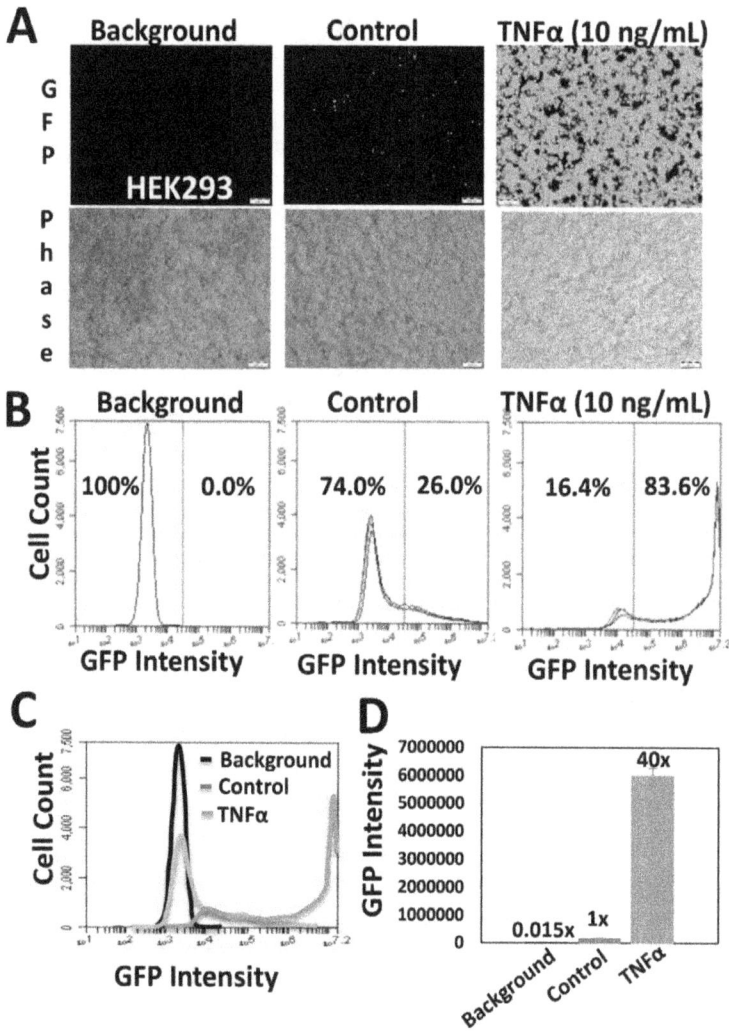

**Figure 6.** High-content quantification of inflammatory response by the flow cytometry. Following transduction with NFκB reporter virus (MOI = 1) or mock transduction (background control) for 24 h, HEK293 cells were switched to serum-free medium in the presence or absence (b) of TNFα (10 ng/mL) for additional 24 h. Response of individual cells was imaged by fluorescence microscopy (A) and further quantified by FACS analysis (B, C). The average response was calculated and graphed (D). The data was presented as mean ± SD, n = 3.

with a laser lamp and a GFP signal detector. As expected, TNFα treatment induced a drastic increase in the GFP intensity over control (**Figure 7A**) with a 16.7-fold increase in GFP signal measured by a microplate reader. To further examine whether our NFκB reporter can be used to monitor the inflammation response in other cell types, we transfected two additional human cell types (U87 and HepG2) with this reporter and quantified their response to TNFα treatment. Similar to HEK293 cells, we observed marked increases (9.2-fold and 18.3-fold over control) in GFP

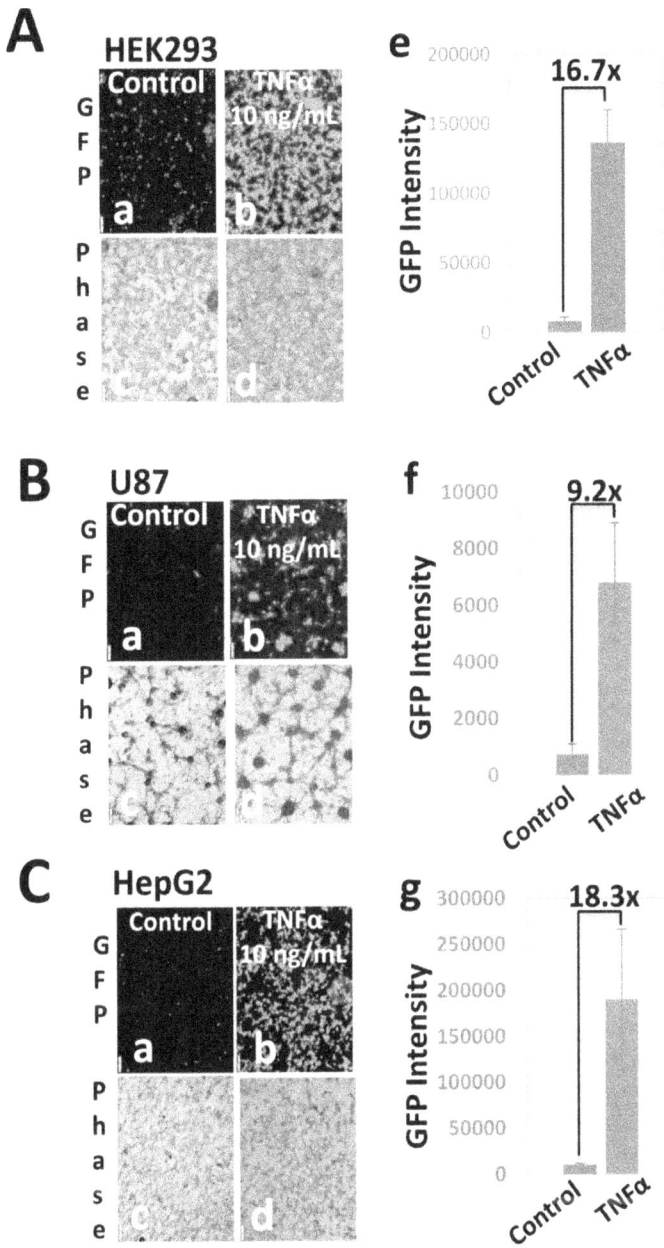

**Figure 7.** Quantification of inflammatory response by microplate reader. Three types of human cells were separately transfected with NF-κB reporter for 24 h and then switched to serum-free complete medium in the presence or absence of TNFα (10 ng/mL) for additional 24 h. The GFP (A–C: a, b) and phase (A–C: c, d) images were taken and the corresponding GFP intensity was further quantified by a microplate reader and graphed (e–g). The data was presented as mean ± SD, n = 3.

intensity in U87 (**Figure 7B**) and HepG2 (**Figure 7C**) cells, respectively. Together, our results validated another convenient approach for signal quantification by using less expensive Microplate reader, which is easily amenable to laboratory automation.

## 4. Discussion

### 4.1. Features of the new genetic circuit

We report the development of a new AAV-based dual-reporter circuit for live monitoring of cell signaling that is fundamental to both physiology and pathology. Our system combines two functionally complementary reporters (Gluc and GFP), which enables HTP applications and laboratory automation. The distinctive feature of this dual-reporter from our previous ones [18, 27] is the introduction of Gluc, which is a secretary form of luciferase and can be easily retrieved from conditioned medium for quantification [17]. Retaining the GFP reporter preserves the capability of both real-time imaging and single cell analysis [34]. Through comprehensive examination and validation using a fluorescence microscope, flow cytometer, and microplate reader, we demonstrated that our new system is robust and provides attractive advantages over existing methods. These advantages include a wide detection range, low background, high sensitivity and better specificity, and multiple gene delivery options.

### 4.2. Genetic circuit better reports cellular response with relevant physiological and pathophysiological significance

The successful development of a genetic circuit requires a sound strategy and an ability to monitor molecular singling in a highly sensitive and specific manner. To study inflammation, we chose the NFκB because it is an important transcription factor in regulating cellular responses with a rapid-activation property [43–45]. In most types of cells, NFκB exists as a dimer in the cytoplasm in an inactive status via interaction with IκB inhibitor proteins. NFκB can be activated by various inflammatory molecules, including TNFα, IL-1β, or bacterial lipopolysaccharides (LPS) [44–46]. Upon binding of these stimuli to their respective receptors, the IκB kinase becomes activated and phosphorylates IκB proteins, which in turn are ubiquitinated and degraded by proteasomes. Once the IκB is degraded, the NFκB complex is free to migrate into nucleus where it binds to its response element and turns on the expression of specific genes that mediate inflammatory responses [47, 48]. Traditionally, methods for the study of inflammation are invasive in nature. Two commonly used methods are Western blot analysis for evaluating IκB activation and RT-PCR for quantifying effector gene expression, both requiring lysis of cells. Additionally, these methods are cumbersome and low in throughput [43, 48]. To overcome these limitations, Lee et al. created a RFP- NFκB reporter cell line, which allows researchers to monitor the NFκB translocation from the cytoplasm to the nucleus [49]. Although this reporter may permit real-time monitoring, its readout is one of the early events of inflammatory signaling, namely the NFκB translocation, rather than the biological endpoint [49]. Due to the oscillation of NFκB, this process may not correlate well with the biological response. Differing from RFP- NFκB reporter, our new system monitors the transcriptional activation, the final step in producing the biological response. Thus, our new system will produce biologically relevant data. Moreover, our dual-reporter format may preserve the signaling history at both individual cell (GFP) and population (Gluc) levels, yielding complimentary data sets as demonstrated in this study (**Figures 3–5**). Most importantly, the endpoint quantification is a direct measurement of NFκB activation in terms of GFP intensity or Gluc activity. Those measurements can be obtained via high content FACS analysis or laboratory automotive microplate reader, in a sensitive and specific manner.

### 4.3. The dual-reporter circuit is applicable to other signaling pathways

Transcriptional activation and control of gene expression is the common focal point that converges on a variety of signaling pathways. Here, we demonstrate that our dual-reporter system can be used to monitor two critical signaling pathways mediated by either by NFκB (inflammation) or AP-1 (cell proliferation and/or differentiation). By the same token, our genetic circuit can be easily modified to report other critical signaling processes such as cancer (P53 or Myc), dyslipidemia (SREBP1 or PPAR), brain development (OCT4 or PAX6) and endocrine function (ER or AR). Hence, this new genetic circuit will have wide applicability and represents a promising platform for studying cell signaling in live cell or animal models.

## 5. Conclusions

We developed and validated a new dual-reporter circuit for real-time monitoring and quantification of inflammatory signaling in various mammalian cells. The new system is readily amenable to noninvasive manipulations allowing high throughput applications and laboratory automation.

## Acknowledgements

We thank Dr. Yan Jiang for critically reviewing the manuscript and helpful comments.

## Author details

Natalie Duong[†], Kevin Curley[†], Mai Anh Do, Daniel Levy and Biao Lu*

*Address all correspondence to: blu2@scu.edu

Department of Bioengineering, School of Engineering, Santa Clara University, Santa Clara, California, USA

[†]These authors contributed equally to this work.

# References

[1]  Tsien RY, Miyawaki A. Seeing the machinery of live cells. Science. 1998;**280**(5371):1954-1955

[2]  Zhang J, Campbell RE, Ting AY, Tsien RY. Creating new fluorescent probes for cell biology. Nature Reviews. Molecular Cell Biology. 2002;**3**(12):906-918

[3]  Chalfie M, Tu Y, Euskirchen G, Ward WW, Prasher DC. Green fluorescent protein as a marker for gene expression. Science. 1994;**263**(5148):802-805

[4]  Kim JB, Urban K, Cochran E, Lee S, Ang A, Rice B, Bata A, Campbell K, Coffee R, Gorodinsky A, et al. Non-invasive detection of a small number of bioluminescent cancer cells in vivo. PLoS One. 2010;**5**(2):e9364

[5]  Kanno A, Yamanaka Y, Hirano H, Umezawa Y, Ozawa T. Cyclic luciferase for real-time sensing of caspase-3 activities in living mammals. Angewandte Chemie (International Ed. in English). 2007;**46**(40):7595-7599

[6]  Li Z, Lee A, Huang M, Chun H, Chung J, Chu P, Hoyt G, Yang P, Rosenberg J, Robbins RC, et al. Imaging survival and function of transplanted cardiac resident stem cells. Journal of the American College of Cardiology. 2009;**53**(14):1229-1240

[7]  Huang M, Chen Z, Hu S, Jia F, Li Z, Hoyt G, Robbins RC, Kay MA, Wu JC. Novel minicircle vector for gene therapy in murine myocardial infarction. Circulation. 2009;**120** (11 Suppl):S230-S237

[8]  Chereau R, Tonnesen J, Nagerl UV. STED microscopy for nanoscale imaging in living brain slices. Methods. 2015;**88**:57-66

[9]  Fernandez-Suarez M, Ting AY. Fluorescent probes for super-resolution imaging in living cells. Nature Reviews. Molecular Cell Biology. 2008;**9**(12):929-943

[10] Jones CT, Catanese MT, Law LM, Khetani SR, Syder AJ, Ploss A, Oh TS, Schoggins JW, MacDonald MR, Bhatia SN, et al. Real-time imaging of hepatitis C virus infection using a fluorescent cell-based reporter system. Nature Biotechnology. 2010;**28**(2):167-171

[11] Bartok E, Bauernfeind F, Khaminets MG, Jakobs C, Monks B, Fitzgerald KA, Latz E, Hornung V. iGLuc: A luciferase-based inflammasome and protease activity reporter. Nature Methods. 2013;**10**(2):147-154

[12] Shaner NC, Campbell RE, Steinbach PA, Giepmans BN, Palmer AE, Tsien RY. Improved monomeric red, orange and yellow fluorescent proteins derived from Discosoma sp. red fluorescent protein. Nature Biotechnology. 2004;**22**(12):1567-1572

[13] Tsien RY. The green fluorescent protein. Annual Review of Biochemistry. 1998;**67**:509-544

[14] Nagai T, Ibata K, Park ES, Kubota M, Mikoshiba K, Miyawaki A. A variant of yellow fluorescent protein with fast and efficient maturation for cell-biological applications. Nature Biotechnology. 2002;**20**(1):87-90

[15] Mezzanotte L, Blankevoort V, Lowik CW, Kaijzel EL. A novel luciferase fusion protein for highly sensitive optical imaging: From single-cell analysis to in vivo whole-body bioluminescence imaging. Analytical and Bioanalytical Chemistry. 2014;406(23):5727-5734

[16] Lorenz WW, McCann RO, Longiaru M, Cormier MJ. Isolation and expression of a cDNA encoding Renilla reniformis luciferase. Proceedings of the National Academy of Sciences of the United States of America. 1991;88(10):4438-4442

[17] Wurdinger T, Badr C, Pike L, de Kleine R, Weissleder R, Breakefield XO, Tannous BA. A secreted luciferase for ex vivo monitoring of in vivo processes. Nature Methods. 2008;5(2): 171-173

[18] Afshari A, Uhde-Stone C, Lu B. Live visualization and quantification of pathway signaling with dual fluorescent and bioluminescent reporters. Biochemical and Biophysical Research Communications. 2014;448(3):281-286

[19] Tay S, Hughey JJ, Lee TK, Lipniacki T, Quake SR, Covert MW. Single-cell NF-kappaB dynamics reveal digital activation and analogue information processing. Nature. 2010;466(7303): 267-271

[20] Stickney Z, Losacco J, McDevitt S, Zhang Z, Lu B. Development of exosome surface display technology in living human cells. Biochemical and Biophysical Research Communications. 2016;472(1):53-59

[21] Badr CE, Hewett JW, Breakefield XO, Tannous BA. A highly sensitive assay for monitoring the secretory pathway and ER stress. PLoS One. 2007;2(6):e571

[22] Koutsoudakis G, Perez-del-Pulgar S, Gonzalez P, Crespo G, Navasa M, Forns X. A Gaussia luciferase cell-based system to assess the infection of cell culture- and serum-derived hepatitis C virus. PLoS One. 2012;7(12):e53254

[23] Canaria CA, Lansford R. Advanced optical imaging in living embryos. Cellular and Molecular Life Sciences. 2010;67(20):3489-3497

[24] Uhde-Stone C, Huang J, Lu B. A robust dual reporter system to visualize and quantify gene expression mediated by transcription activator-like effectors. Biological Procedures Online. 2012;14(1):8

[25] Afshari A, Uhde-Stone C, Lu B. A cooled CCD camera-based protocol provides an effective solution for in vitro monitoring of luciferase. Biochemical and Biophysical Research Communications. 2015;458(3):543-548

[26] Tannous BA. Gaussia luciferase reporter assay for monitoring biological processes in culture and in vivo. Nature Protocols. 2009;4(4):582-591

[27] Zhang Z, Stickney Z, Duong N, Curley K, Lu B. AAV-based dual-reporter circuit for monitoring cell signaling in living human cells. Journal of Biological Engineering. 2017;11:18

[28] Tannous BA, Kim DE, Fernandez JL, Weissleder R, Breakefield XO. Codon-optimized Gaussia luciferase cDNA for mammalian gene expression in culture and in vivo. Molecular Therapy. 2005;11(3):435-443

[29] Uhde-Stone C, Sarkar N, Antes T, Otoc N, Kim Y, Jiang YJ, Lu B. A TALEN-based strategy for efficient bi-allelic miRNA ablation in human cells. RNA. 2014;**20**(6):948-955

[30] Grimm D, Lee JS, Wang L, Desai T, Akache B, Storm TA, Kay MA. In vitro and in vivo gene therapy vector evolution via multispecies interbreeding and retargeting of adeno-associated viruses. Journal of Virology. 2008;**82**(12):5887-5911

[31] Potter M, Lins B, Mietzsch M, Heilbronn R, Van Vliet K, Chipman P, Agbandje-McKenna M, Cleaver BD, Clement N, Byrne BJ, et al. A simplified purification protocol for recombinant adeno-associated virus vectors. Molecular therapy. Methods & Clinical Development. 2014;**1**:14034

[32] Zolotukhin S, Potter M, Zolotukhin I, Sakai Y, Loiler S, Fraites TJ, Chiodo VA, Phillipsberg T, Muzyczka N, Hauswirth WW, et al. Production and purification of serotype 1, 2, and 5 recombinant adeno-associated viral vectors. Methods. 2002;**28**(2):158-167

[33] Sengupta R, Mukherjee C, Sarkar N, Sun Z, Lesnik J, Huang J, Lu B. An optimized protocol for packaging Pseudotyped integrase defective lentivirus. Biological Procedures Online; **2017, 18**:14

[34] Wiedenmann J, Oswald F, Nienhaus GU. Fluorescent proteins for live cell imaging: Opportunities, limitations, and challenges. IUBMB Life. 2009;**61**(11):1029-1042

[35] Kellogg RA, Tay S. Noise facilitates transcriptional control under dynamic inputs. Cell. 2015;**160**(3):381-392

[36] Lu B, Lu Y, Moser AH, Shigenaga JK, Grunfeld C, Feingold KR. LPS and proinflammatory cytokines decrease lipin-1 in mouse adipose tissue and 3T3-L1 adipocytes. American Journal of Physiology. Endocrinology and Metabolism. 2008;**295**(6):E1502-E1509

[37] Lu B, Moser A, Shigenaga JK, Grunfeld C, Feingold KR. The acute phase response stimulates the expression of angiopoietin like protein 4. Biochemical and Biophysical Research Communications. 2010;**391**(4):1737-1741

[38] Lu B, Moser AH, Shigenaga JK, Feingold KR, Grunfeld C. Type II nuclear hormone receptors, coactivator, and target gene repression in adipose tissue in the acute-phase response. Journal of Lipid Research. 2006;**47**(10):2179-2190

[39] Johnston SR, Lu B, Scott GK, Kushner PJ, Smith IE, Dowsett M, Benz CC. Increased activator protein-1 DNA binding and c-Jun NH2-terminal kinase activity in human breast tumors with acquired tamoxifen resistance. Clinical Cancer Research. 1999;**5**(2):251-256

[40] Burger C, Gorbatyuk OS, Velardo MJ, Peden CS, Williams P, Zolotukhin S, Reier PJ, Mandel RJ, Muzyczka N. Recombinant AAV viral vectors pseudotyped with viral capsids from serotypes 1, 2, and 5 display differential efficiency and cell tropism after delivery to different regions of the central nervous system. Molecular Therapy. 2004;**10**(2):302-317

[41] Xiao X, Li J, Samulski RJ. Efficient long-term gene transfer into muscle tissue of immunocompetent mice by adeno-associated virus vector. Journal of Virology. 1996;**70**(11):8098-8108

[42] Zeltner N, Kohlbrenner E, Clement N, Weber T, Linden RM. Near-perfect infectivity of wild-type AAV as benchmark for infectivity of recombinant AAV vectors. Gene Therapy. 2010;**17**(7):872-879

[43] Hoffmann A, Baltimore D. Circuitry of nuclear factor kappaB signaling. Immunological Reviews. 2006;**210**:171-186

[44] Cheong R, Bergmann A, Werner SL, Regal J, Hoffmann A, Levchenko A. Transient IkappaB kinase activity mediates temporal NF-kappaB dynamics in response to a wide range of tumor necrosis factor-alpha doses. The Journal of Biological Chemistry. 2006;**281**(5):2945-2950

[45] Covert MW, Leung TH, Gaston JE, Baltimore D. Achieving stability of lipopolysaccharide-induced NF-kappaB activation. Science. 2005;**309**(5742):1854-1857

[46] Muzio M, Ni J, Feng P, Dixit VM. IRAK (Pelle) family member IRAK-2 and MyD88 as proximal mediators of IL-1 signaling. Science. 1997;**278**(5343):1612-1615

[47] Nelson DE, Ihekwaba AE, Elliott M, Johnson JR, Gibney CA, Foreman BE, Nelson G, See V, Horton CA, Spiller DG, et al. Oscillations in NF-kappaB signaling control the dynamics of gene expression. Science. 2004;**306**(5696):704-708

[48] Hoffmann A, Levchenko A, Scott ML, Baltimore D. The IkappaB-NF-kappaB signaling module: Temporal control and selective gene activation. Science. 2002;**298**(5596):1241-1245

[49] Lee TK, Denny EM, Sanghvi JC, Gaston JE, Maynard ND, Hughey JJ, Covert MW. A noisy paracrine signal determines the cellular NF-kappaB response to lipopolysaccharide. *Science Signaling*. 2009;**2**(93):ra65

# Chapter 4

# The Signaling Nature of Cellular Metabolism: The Hypoxia Signaling

Zsolt Fabian

Additional information is available at the end of the chapter

http://dx.doi.org/10.5772/intechopen.79952

### Abstract

Identification of the hypoxia-inducible factors (HIFs) as core players of the transcriptional response to hypoxia transformed our understanding of the mechanism underpinning the hypoxic response at the molecular level and led to discoveries on the role of metabolism in cell signaling alike. It has now become clear that HIFs act in the heart of a pathway where oxygen may be considered as a signaling entity recognized by molecular sensors conveying the oxygen signal to the transcriptional regulator HIFs as distal effectors. The pathway is under multiple levels of regulatory control shaping the cellular response to hypoxia and gives hypoxia signaling an intricate and dynamic activity profile. These include regulatory mechanisms within the HIF pathway as well as diverse interplay with other metabolic and signaling pathways of critical cellular functions. The emerging model reflects a multi-level regulatory network that apparently affects all aspects of cell physiology.

**Keywords:** hypoxia, hypoxia-inducible factors, prolyl hydroxylases

## 1. Introduction

The development of molecular machineries capable of utilizing atmospheric oxygen for bioenergetic purposes was a key event in the evolution of life on Earth. This, along with other processes like compartmentalization, allowed eukaryotic organisms to substantially enhance metabolic efficiency. The accompanying development of a range of biochemical processes provided the bioenergetic capacity to permit the evolution of more complex life forms of metazoans [1]. In parallel, the high degree of dependence of a constant oxygen supply to maintain metabolic homeostasis provoked the evolution of counter measures termed the hypoxia pathway [2].

## 2. The hypoxia pathway

The critical dependence on oxygen for metabolic homeostasis and survival led to the early evolution of a molecular mechanism that enabled cells, tissues, and organisms to adapt to hypoxia. This adaptive response is primarily orchestrated by a family of transcription factors termed hypoxia-inducible factors (HIFs) [3]. In mammals, three members of the helix-loop-helix-type HIF family have been identified to date (HIF-1, HIF-2, and HIF-3) of which the prototype is HIF-1 (**Figure 1**). The active HIF-1 is composed of discrete alpha and beta subunits (HIF-1α and HIF-1β, respectively) both of which are ubiquitously expressed in human tissues, whereas HIF-2α and HIF-3α are selectively expressed in certain cell types [4, 5]. Unlike HIF-β that is stably expressed in the cells, HIF-α subunits are continuously degraded by the 26S proteasome under normoxic conditions. This mechanism prevents the formation and activity of HIF heterodimers in sufficiently oxygenized cells and the launch of their hypoxia-adaptive genetic program. In hypoxia, however, HIF-α subunits escape from the constitutive degradation, become stabilized in the cytoplasm, dimerize with HIF-1β and the nuclear heterodimers rearrange gene expression pattern of the hypoxic cell. This primarily, but not exclusively, includes induction of genes that mediate the switch from oxygen-dependent to anaerobic metabolism.

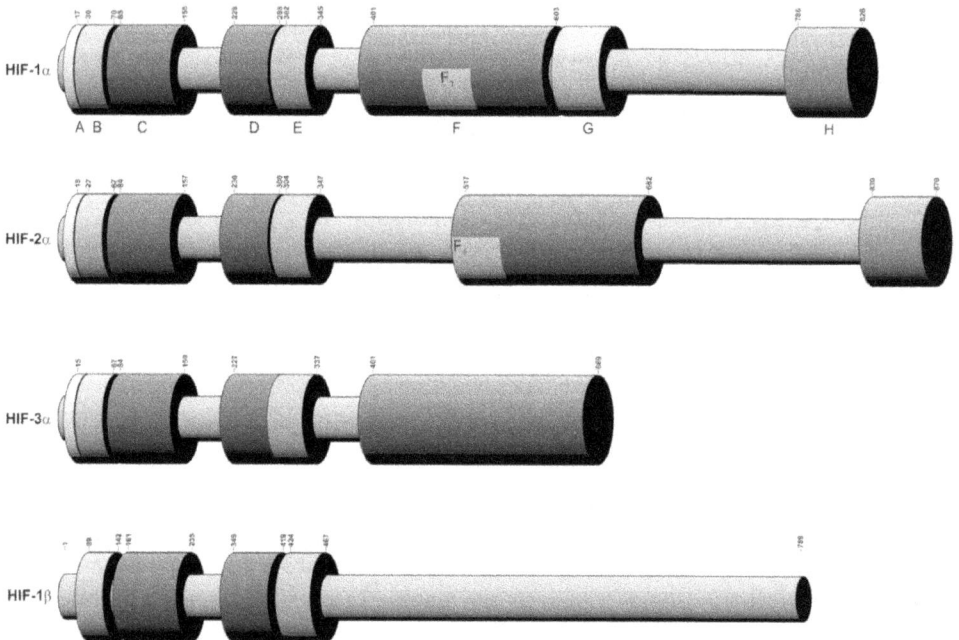

**Figure 1.** Domain structure of HIF polypeptides A: DNA-binding domain; B: basic helix-loop-helix domain; C: Per/Arnt/Sim (PAS) A domain; D: PAS B domain; E: PAC motif; F: oxygen-dependent degradation domain; $F_{1:}$ N-terminal transactivation domain; G: ERK target domain; H: C-terminal transactivation domain. Amino acid positions indicated are based on current Uniprot.

The molecular background of normoxic degradation of HIF-α was first elucidated in 2001 [6, 7] (**Figure 2**). It turned out that its continuous proteasomal elimination is triggered by the oxygen-dependent hydroxylation mediated by a family of prolyl-4-hydroxylases reminiscent of procollagen prolyl hydroxylases that had long been known at the time. To date, three HIF-regulating prolyl-4-hydroxylases (also known as PHD1, PHD2, and PHD3) have been identified in mammalian cells [8]. They utilize molecular oxygen, ascorbic acid, iron, and the tricarboxylic acid (TCA) cycle intermediate α-ketoglutarate as co-factors and co-substrates to hydroxylate the HIF-α subunits at conserved prolyl residues [6, 9]. In HIF-1α, these are proline 402 and 564 and their hydroxylation increases the affinity of the polypeptide to the von Hippel-Lindau protein (pVHL), the substrate recognition component of the E3 ubiquitin ligase complex of Elongin-B and -C, Cul2, and Rbx1 [10]. This leads to pVHL-mediated ubiquitylation of lysine residues (lysine 532, 538, and 547 in case of HIF-1α) within the so-called oxygen-dependent degradation domain that renders the polypeptide for constitutive proteasomal degradation. In hypoxia, this hydroxylation activity is reduced due to the lack of available oxygen, resulting in stabilization of the HIF-α subunits. In addition to the PHD-mediated post-translational modifications, a second level of hydroxylation-dependent regulation of HIF-α has also been

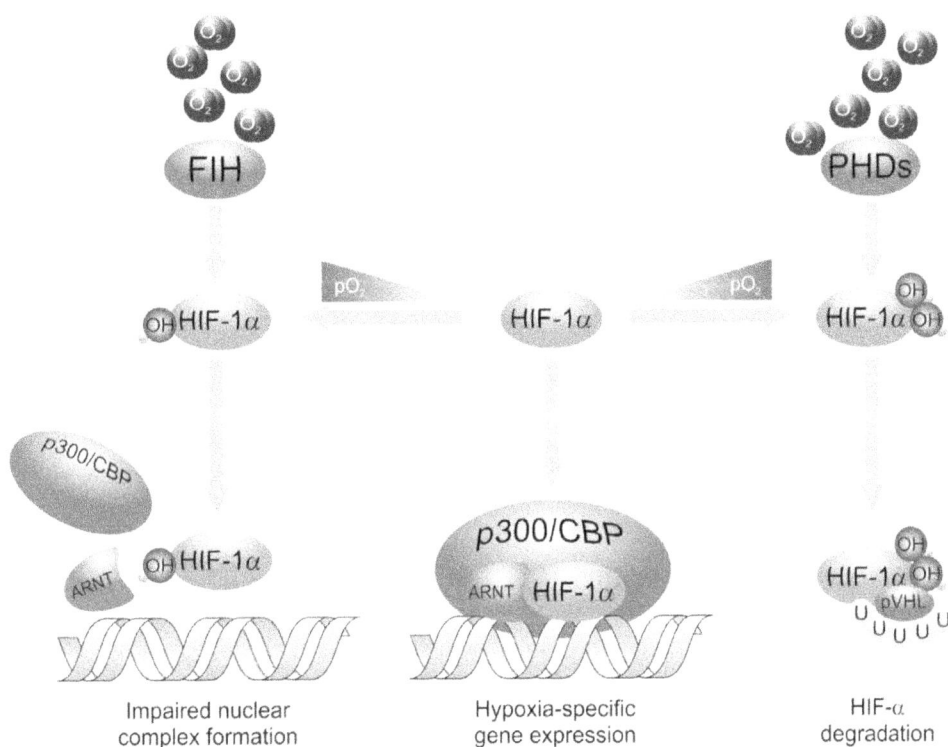

**Figure 2.** Oxygen sensing by hydroxylases. Abbreviations: FIH: factor inhibiting HIF; PHDs: prolyl-4-hydroxylases; HIF: hypoxia-inducible factor; ARNT: aryl hydrocarbon receptor nuclear translocator; pVHL: von Hippel-Lindau ubiquitin ligase; OH: hydroxylation; U: ubiquitylation; p300/CBP: transcriptional co-activators of HIF heterodimers.

discovered. This post-translational modification is mediated by the asparagine hydroxylase termed factor inhibiting HIF (FIH). Unlike PHDs, however, the FIH-mediated hydroxylation is believed to primarily prevent HIF's interaction with transcriptional co-activators like the p300/ CBP (**Figure 2**) [11–13].

In the hypoxia pathway, oxygen may be considered the "ligand" for oxygen sensor prolyl-4-hydroxylases that principally govern HIF activity. This, eventually, culminates in the launch of a complex adaptive program that fundamentally affects cellular homeostasis via metabolic switch from the oxygen-dependent oxidative phosphorylation to glycolysis, increased angiogenesis and enhanced erythropoiesis. Thus, it is not surprising that a range of additional inputs, including feedback loops and multiple cross talks with other signaling pathways, shape the spatial and temporal nature of the ultimate response to oxygen depletion. These interactions form a metabolic signaling network that confers a dynamic profile and a high degree of complexity upon the hypoxic response.

## 3. Regulatory measures in hypoxia signaling

### 3.1. Supracellular signaling

In principle, adaptation to hypoxia may involve two directions of counter measures; reduction of oxygen consumption and increase of oxygen supply. In multicellular organisms, the latter one requires coordinated supracellular, multi-organ measures governed by HIF-inducible genes including elevated erythropoiesis and angiogenesis. At a systemic level, hypoxia-activated HIFs induce erythropoietin (EPO) expression in liver and interstitial kidney cells that, subsequently, triggers erythropoiesis in the bone marrow [14, 15]. This elevated red blood cell production, however, requires increased iron supply of bone marrow erythroblasts regulated by, at least in part, the hepatocyte-specific iron homeostasis regulator hepcidin. This short peptide is believed to be responsible for inhibiting the iron release and absorption from macrophages and intestinal epithelial cells, respectively, by binding the only known cellular iron exporter ferroportin [16]. Upregulated HIF-driven erythropoiesis provokes repression of the hepatic hepcidin-encoding gene, although the identity of the soluble mediator of this effect is yet to be confirmed [17–20]. The drop of serum hepcidin upon hypoxia, eventually, results in elevated iron release from the intestinal epithelium supplying the increased iron demand of expanded erythropoiesis [21].

When hypoxia develops locally, sheer increase of the oxygen transport capacity may not be sufficient to elevate the oxygen supply of hypoxic tissues. In these conditions, hypoxia is accompanied by angiogenesis representing another tissue-level negative feedback loop of hypoxia signaling. This arm of the regulation is mediated by the key angiogenesis-regulating growth factor termed vascular endothelial growth factor A (VEGF-A). Similar to *EPO, VEGFA*, the key determinant of survival and proliferation of endothelial cells upon embryonic vasculogenesis, is another common target of HIF-1 and -2 [22, 23]. In addition, high levels of VEGF-A expressed by hypoxic stromal or tumor cells regulate endothelial cells metabolism by biasing it toward glycolysis via induction of isoform 3 of 6-phospho-fructo-2-kinase/fructose-2,6-biphosphatase [24]. Increased glycolysis not only supports endothelial cell survival under hypoxic conditions but also triggers vessel sprouting, further representing a complementary mechanism of VEGF-mediated angiogenesis [25]. Elevated oxygen and iron levels, consequently, provide prolyl-4-hydroxylases increased supply of their co-substrates completing the supracellular regulatory loop of hypoxia signaling [26].

## 3.2. Intracellular metabolic signaling in hypoxia

At the cellular level, HIFs reprogram metabolism directly targeting a cluster of metabolic enzyme-coding genes [27]. Their prototype is pyruvate dehydrogenase kinase-1 (PDK-1) which phosphorylates pyruvate dehydrogenase (PDH), the enzyme that supplies TCA cycle with acetyl-coenzyme A [28]. PDK-1 mediates inactivating phosphorylation of PDH that shuts down the TCA cycle due to shortage of acetyl-coenzyme A. This leads to fundamental changes in mitochondrial functions including the accumulation of TCA cycle intermediates [29]. Since HIF-regulating prolyl-4-hydroxylases utilize $\alpha$-ketoglutarate and produce succinate during their catalytic activity, one can speculate that accumulation of the latter one blocks catalytic activity of prolyl-4-hydroxylases [30]. Indeed, it was found that loss-of-function mutations of succinate dehydrogenase, the TCA cycle enzyme that converts succinate into fumarate, block PHDs and, consequently, stabilize HIF-$\alpha$ subunits [31]. In addition, it has also been demonstrated that this effect is, primarily, mediated by the accumulation of succinate (**Figure 3**) [32].

Besides their direct metabolic target genes, HIFs also regulate the hypoxia pathway sensor PHDs indirectly through the HIF-inducible microRNA-210 (miR-210)-mediated silencing of the glycerol-3-phosphate dehydrogenase 1 like protein (GPD1-L). GPD1-L has a similar enzymatic activity to that of the mitochondrial glycerol-3-phosphate dehydrogenase and catalyzes the redox conversion of glycerol-3-phosphate (G3P) to dihydroxyacetone-phosphate [33]. Although the mechanism behind the connection is still not clear, the miR-210-mediated downregulation of GPD1-L is accompanied by increased PHD-mediated HIF-1$\alpha$ degradation [34]. Since decreased enzymatic GPD1-L activity results in increased G3P levels, upregulated glycolysis may contribute to the redistribution of available oxygen from mitochondria to PHDs and, thus, represents the link between miR-210 and the restoration of PHD activity. This hypothesis is further supported by the observation that inhibition of the mitochondrial respiration by nitric oxide is followed by redistribution of $O_2$ and inactivation of HIFs [35].

The concept of metabolite-mediated regulation of PHDs is further supported by the observation that, in hypoxia, another microRNA, miR-183, targets isocitrate dehydrogenase, the TCA cycle mediator that produces $\alpha$-ketoglutarate from isocitrate. Although the mechanism of its hypoxic upregulation is yet to be determined, the miR-183-mediated blockade of $\alpha$-ketoglutarate production exploits the $\alpha$-ketoglutarate-dependent nature of prolyl-4-hydroxylases and promotes stabilization of HIF-$\alpha$ via inhibition of PHDs [36]. Thus, PHDs not only act as oxygen sensors but can also integrate metabolic stimuli forming synergistic metabolic positive feedback loops within hypoxia signaling [37].

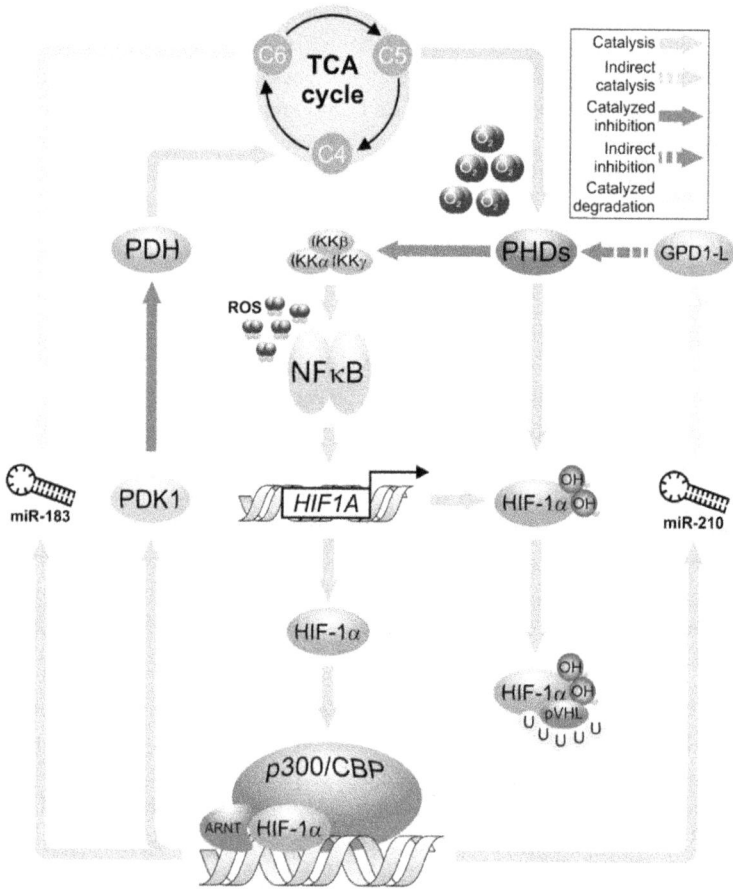

**Figure 3.** Interplay of the hypoxia and metabolic signaling. Abbreviations: TCA cycle: tricarboxylic acid cycle; C4, C5, and C6 represent the 4, 5, and 6 carbon metabolites of the TCA cycle, respectively; IKKα, IKKβ, and IKKγ are the Inhibitory kappa-B kinase alpha, beta, and gamma subunits, respectively; PDH: pyruvate dehydrogenase; PDK1: PDH kinase 1; GPD1-L: Glycerol-3-phosphate dehydrogenase 1-like protein; PHDs: prolyl-4-hydroxylases; NF-κB: nuclear factor kappa-B; HIF: hypoxia-inducible factor; ARNT: aryl hydrocarbon receptor nuclear translocator; pVHL: von Hippel–Lindau ubiquitin ligase; OH: hydroxylation; U: ubiquitylation; p300/CBP: transcriptional coactivators of HIF heterodimers; miR-183 and miR-210 are microRNA-183 and microRNA-210, respectively.

### 3.3. Transcriptional feedbacks

HIFs directly induce genes with wide range of functions including both *PHD2* and *PHD3* (**Figure 4**). Although HIF-mediated transactivation of *PHDs* resembles a canonical, direct negative feedback arm within the PHD-HIF axis, it may have more complex functions [38, 39]. HIFs can only transactivate their targets upon oxygen depletion so one can speculate that, induction of the oxygen-dependent PHDs is useless under hypoxia. In fact, however, despite their oxygen dependency, which prevents them from functioning under hypoxic conditions, experimental data indicate that enzymatic activity of both PHD2 and PHD3 remains detectable in hypoxic

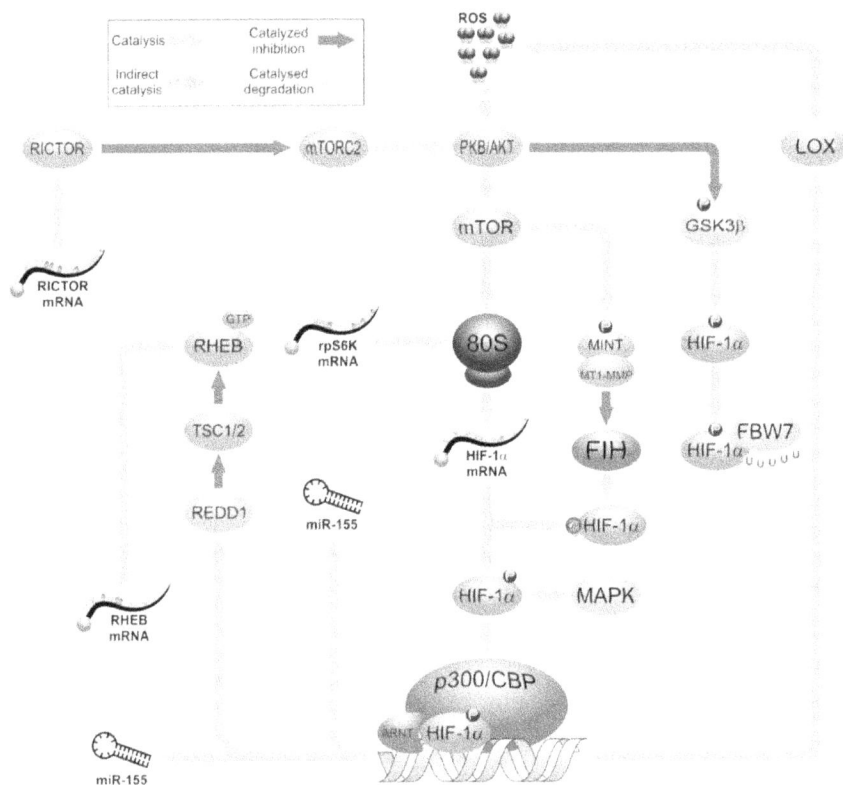

**Figure 4.** Cross talk between the hypoxia, anabolic, mitogen, and inflammatory signaling pathways. Abbreviations: ROS: reactive oxygen species; RICTOR: rapamycin-insensitive companion of mTOR; mTORC2: mammalian target of rapamycin complex 2; PKB/AKT: protein kinase B; mTOR: mammalian target of rapamycin; LOX: Lysyl oxidase; GSK3β: glycogen synthase kinase-3 beta; RHEB: Ras homolog mTOR complex 1-binding protein; 80S: mature eukaryotic ribosome; MINT: Amyloid beta precursor protein binding family A, member 3; MT1-MMP: membrane-type 1 matrix metalloproteinase; TSC1/2: Tuberous Sclerosis complex 1/2; REDD1: DNA damage-inducible transcript 4; PHDs: prolyl-4-hydroxylases; FIH: factor inhibiting HIF; MAPK: mitogen activated protein kinase; HIF: hypoxia-inducible factor; ARNT: aryl hydrocarbon receptor nuclear translocator; FBW7: F-Box and WD-40 domain-containing protein 7; OH: hydroxylation; U: ubiquitylation; p300/CBP: transcriptional coactivators of HIF heterodimers; miR-155: microRNA-155.

cells maintaining reactivity of the HIF system for further hypoxic insults [39]. Thus, HIF-mediated induction of *PHDs* in conjunction with their metabolic effects, possibly, functions as a mechanism responsible for resetting hypoxia signaling to a new steady state at lower oxygen levels.

The HIF-target miR-210 also plays multiple, apparently opposing, roles in the regulation of the hypoxia pathway. It not only indirectly facilitates HIF activity by targeting GPD1-L but also silences MYC antagonist MNT, a member of the MYC/MAD/MYX transcription factor family [40]. This downregulation deliberates MYC-mediated induction of genes like those involved in the resolution of HIF-induced cell cycle arrest or metabolic switch via PDK1 illustrating the complexity of the HIF-provoked signaling responses. While, through the former action, MYC counteracts hypoxia signaling, MYC-mediated induction of PDK1 synergizes with HIF activity [41]. Since MYC has also been reported to support HIF-1α directly by interfering with the VHL-dependent degradation of HIF-1α, data strongly suggest the existence of a MYC-mediated feedforward loop in the HIF pathway [42]. In return, HIF counteracts MYC by various underlying mechanisms including the induction of MXI1, another MYC antagonist, competition with MYC for promoter binding or promoting its proteasomal degradation [43, 44]. Since the opposing effects of the HIF/MYC interaction in the regulation of cell cycle and metabolism may reflect differences of the experimental models used, the biological relevance of the hypoxia-inducible miR-210-promoted MYC functions requires further investigations. Additional targets of miR-210 like the mitochondrial iron-sulfur cluster scaffold protein or transferrin suggest the potential signal integration role of miR-210 in hypoxia signaling [45, 46].

Besides miR-210, the HIF-inducible miR-155 represents another level of microRNA-mediated transcriptional regulation of the hypoxia pathway. Upon a hypoxic insult, it shapes dynamics of the HIF-response by facilitating the RISC-mediated degradation of the HIF-1α transcript [47]. Intriguingly, the miR-155-mediated direct silencing of HIF-α expression not only illustrates an isoform-specific resolution of hypoxia signaling upon prolonged hypoxia but also resembles the HIF-mediated induction of PHDs and might ensure the cellular reactivity to hypoxia at lower $pO_2$ levels.

An additional form of the transcriptional regulation of hypoxia signaling is mediated by the *HIF3A-encoded* isoform termed inhibitory PAS domain protein (IPAS). IPAS is an alternative splicing product of the *HIF3A* locus and generates a polypeptide that lacks the C-terminal transactivation domains of HIF-1 and -2α (**Figure 1**) [48]. As such, it functions as a dominant negative regulator of HIFs by competing with HIF-1β [49]. The IPAS-specific splicing product is hypoxia-inducible and, at last in part, is under the control of a HIF-1-specific hypoxia-response element representing one of the classic negative feedback loops of the hypoxia pathway [48, 50]. Interestingly, the IPAS-specific mRNA splicing also takes place in the absence of the HIF-1-binding site of the IPAS promoter suggesting the existence of HIF-independent factors involved in the expression of IPAS [50]. The uncoupled nature of IPAS expression and IPAS mRNA splicing underpins the presence of an additional control layer in the IPAS-mediated HIF regulation. Indeed, since normoxic expression of IPAS is, apparently, restricted to corneal epithelial cells and some neuronal elements in mice, the HIF-independent control mechanisms may contribute to the tissue-specific nature of the IPAS-mediated regulation of hypoxia signaling.

## 4. Cross talks

### 4.1. Cross talk through HIF-1β

Due to the fundamental role of oxygen in the cellular homeostasis, the hypoxic insult requires counter measures that rely on tight coordination of the full spectrum of cellular functions. As part of this, extensive interplays exist between the primary hypoxia sensing PHD-HIF axis and distinct molecular machineries involved in cellular functions like catabolism, cell cycle, or cellular defense mechanisms. One of the mediators of these interactions is the β subunit of HIF heterodimers also known as the aryl hydrocarbon nuclear translocator (ARNT).

Besides its critical role in the formation of active HIFs, the constitutively expressed HIF-1β, as its alias indicates, is also the partner of the aryl hydrocarbon receptor (AhR), a transcription factor that targets genes involved in the biotransformation of xenobiotics [51, 52]. The class I bHLH/PAS protein family member AhR is ubiquitously expressed and activated by various endo- and exogenous ligands. In its inactive state, it forms heterodimers with repressor proteins, like the heat shock protein 90, in the cytoplasm [53]. Upon ligand binding, its nuclear localization signal becomes exposed and, following the consequent nuclear translocation, AhR dimerizes with HIF-1β [54]. The class II bHLH/PAS HIF-1β is essential for the AhR-mediated induction of genes with 5′-TNGCGTG-3′ sequences (also known as xenobiotic response element [XRE]) present in their promoters [55–57]. These include phase I and II detoxifying enzymes like the cytochrome P450 (CYP1A1) and UDP-glucuronosyltransferase 1 isoforms, respectively [58].

Due to their shared partner, it was proposed that activation of hypoxia signaling affects AhR-mediated responses and *vice versa*, AhR-mediated engagement of HIF-1β attenuates hypoxia responses. Indeed, in hypoxic cells exposed to 2,3,7,8-tetrachlorodibenzo-*p*-dioxin (TCDD), the prototype ligand for AhR, the AhR-mediated CYP1A1 expression was found reduced. In contrast, TCDD treatment of cells inhibited HIF-mediated transcriptional responses indicating the cross talk between the AhR and hypoxia signaling [59]. Compromised formation of the AhR/HIF-1β or HIF-1α/HIF-1β heterodimers, in the presence of hypoxia or AhR ligands, respectively, indicates that, in human cells, HIF-1β acts as the limiting factor of the dimer formation [60]. Experimental data also indicates that HIF-1α has higher affinity to HIF-1β than that of the AhR, suggesting that human cells experience the oxygen depleted milieu as a potentially more harmful environmental stimulus than that of the presence of xenobiotics [61]. In the case of genes under the regulation of both HRE and XRE sequences, cross talk between the hypoxia and AhR pathways is even more complex as simultaneous presence of AhR ligands and hypoxia was found to be rather additive, reflecting the cell-type or stimulus-dependent nature of these responses [62].

## 4.2. Cross talk between the hypoxic and anabolic signaling

Hypoxia is one of the strongest stimuli of autophagy that is considered as an indicator of depleted ATP pools of hypoxic cells. In concert, the HIF-orchestrated adaptive program involves downregulation of catabolic pathways like the one regulated by the mammalian target of rapamycin (mTOR) (**Figure 4**). The involvement of mTOR in the regulation of HIF-1 was first suggested by independent studies on the oncogene-related activation of VEGF [63–66]. Unlike PHDs, mTOR enhances HIF-1-mediated transcriptional activity without affecting its degradation rate [64, 67]. mTOR alters the protein expression pattern of hypoxic cells, at least in part, via phosphorylation of the eukaryotic initiation factor 4E-binding protein 1 (eIF4E-BP1), a suppressor of the 5′ CAP-dependent translation [68]. These observations together with findings on the negative effects of rapamycin on HIF-1 suggest that mTOR enhances HIF-α mRNA translation [69]. This concept is also supported by the observation that downregulation of the mTOR complex 2 (mTORC2), a redox-sensitive activator of the PKB/AKT pathway, leads to decreased abundance of the HIF-2α transcripts in the polysomal fractions [70]. Current data indicate that the dominant upstream regulator of the hypoxia-related mTOR activity is protein kinase B (PKB/AKT) [67]. Hypoxic activation of PKB/AKT that, at least in part, depends on reactive oxygen species (ROS), was shown to regulate PHD activity and promote stabilization of HIFs [71–73]. Intriguingly, HIF is actively involved in the generation of ROS under hypoxia by inducing lysyl oxidase (LOX) [67]. *LOX* encodes a copper-dependent amine oxidase that catalyzes cross-linking of collagen and elastin in the extracellular matrix while producing hydrogen peroxide ($H_2O_2$). Current data indicate that, following its HIF-dependent upregulation in hypoxia, LOX-generated $H_2O_2$ activates the PKB/AKT-mTOR axis resulting in the upregulation of HIF-1α translation illustrating a positive feedback loop between mTOR and HIF [67]. It is noteworthy that HIF-mediated induction of PHDs, thus, may not only play a role in resetting the hypoxia pathway at a lower oxygen tension but also represents a limiting step of the mTOR-mediated enhancement of HIF-1α translation [74].

PKB/AKT has also been shown to affect the proteasomal degradation of HIF-1α through glycogen synthase kinase 3β (GSK3β) [75]. GSK3β-mediated phosphorylation of HIF-1α facilitates its binding to FBW7, an E3 ubiquitin ligase, which recognizes GSK3β-phosphorylated proteins and targets them for proteasomal degradation [76]. Since inactivating phosphorylation of GSK3β is primarily mediated by PKB/AKT, activation of the PKB/AKT pathway not only influences the translational rate of HIF-1α via mTOR but also mimics the effect of hypoxia via inhibition of proteasomal degradation of the HIF-1α polypeptide.

To make the picture even more colorful, mTOR also phosphorylates MINT3, a membrane-type matrix metalloproteinase (MT-MMPs) regulator, at its threonine 5/serine 7 residues [77]. This modification promotes binding of MINT3 to FIH-1 leading to the inactivation of the latter [78]. By sequestering the HIF-1 suppressor FIH-1 to the Golgi membrane in cooperation with the MT1-MMP, the mTOR/MINT3/MT1-MMP axis can also support transcriptional activity of HIF-1 independently of its translational rate. Interestingly, in renal cell carcinoma, MT1-MMP has been found to be a target gene for HIF-2 raising the question if the mTOR-regulated MINT3/MT1-MMP/FIH-1-mediated positive feedback loop is a general mechanism in the regulation of HIFs [79]. Although its biological relevance is yet to be determined, it is noteworthy that mTOR has also been reported to associate with HIF-1α via the mTOR complex 1 member RAPTOR and a putative TOR motif within the HIF-1α polypeptide [66]. Since mTOR is a serine/threonine kinase and one of the known posttranslational modifications that favors HIF-1α transcriptional activity is phosphorylation, the possibility that this co-localization also supports the effect of mTOR on HIF-1 via direct phosphorylation cannot be excluded but is yet to be confirmed [66, 80].

Eventually, rapamycin-sensitive upregulation of HIF-1 supports the induction of a wide range of HIF-1 targets, of which many have been found to form feedback loops via regulation of the hypoxia-related activity of mTOR. These include REDD1 that activates the tuberous sclerosis complex 1/2 (TSC1/2) [81]. The TSC1/2 possesses GTPase-activating function that renders the mTOR activator RHEB inactive [82]. BNIP3, another known HIF-1 target, also facilitates the accumulation of the GDP-bound form of RHEB and the consequent downregulation of mTOR in hypoxia [83]. In addition, the HIF-1-inducible miR-155 also targets elements of the mTOR pathway including RHEB, the mTORC2 member RICTOR and the mTOR effector ribosomal protein S6KB2 [84]. Downregulation of these targets, seemingly, complements the effect of REDD1 and BNIP3 and may contribute to the limitation of mTOR signaling in hypoxia.

## 4.3. Interplay with the imflammatory and mitogen signaling pathways

Besides its role in the mTOR-mediated HIF regulation discussed above, hypoxia-inducible miR-155 is one of the identified measures directly targeting HIF-1α mRNA, indicating its pivotal role in the regulation of the HIF pathway. Sequence analyses revealed that, besides the HIF respon-sive element, NF-κB consensus sequences are also present in the miR-155 promoter indicating the capacity of NF-κB-mediated stimuli to influence the HIF pathway via miR-155 [47]. Studies on the proposed link between hypoxia and inflammation revealed that NF-κB can also induce *HIF1A* via evolutionary conserved consensus binding sites identified in the *HIF1A* promoter [72, 85–87]. Since this induction is not sufficient for the accumulation of HIF-1α in the absence of hypoxia, current data suggest that the canonical NF-κB pathway is rather for pre-setting the HIF-1α mRNA level according to the redox state and inflammatory cytokine composition of the extracellular milieu [88]. This concept is further supported by the observation that NF-κB activity can be both up- and down-regulated upon inhibition of prolyl-4-hydroxylases depending on the NF-κB stimulus received [89].

Intriguingly, besides *HIF1A*, NF-κB transactivates *ARNT* as well, leading to enhanced formation of HIF-1β:HIF-2α that attenuates the proteasomal degradation of the latter. Considering the cell-type specific expression pattern of HIF-2α, the NF-κB-mediated induction of *ARNT* may represent a tissue-specific arm of the NF-κB-governed regulation of hypoxia signaling [90]. The interplay, however, is apparently bidirectional and the hypoxic signal can also be conveyed to the inflam-matory pathway. Under normoxic conditions, PHDs inhibit the I kappa B kinase (IKK) complex attenuating the dissociation of inhibitory kappa B (IκB) from NF-κB [85, 88, 91]. In the absence of oxygen, however, the PHD-mediated blockade of IKK is resolved, the IKK complex becomes active leading to the phosphorylation of IκB and release of sequestered NF-κB subunits. The con-sequent formation of active NF-κB heterodimers culminates in moderate upregulation of the basal NF-κB activity that is believed to potentiate NF-κB responsiveness to cytokines like the tumor necrosis factor-alpha (TNF-α) or reactive oxygen species, stimuli typically accompanying inflam-matory conditions [72, 92–95]. Thus, the interaction between the NF-κB and HIF pathways well illustrates the close pathophysiologic connection between hypoxia and inflammation and allows the cell to integrate inflammatory stimuli in the adaptive response under hypoxic conditions.

Anabolic extracellular signals that activate the mTOR pathway often diverge and activate the ERK signaling cascade as well, raising the question if ERK and hypoxia signaling interplay. Experimental data indicate that the HIF-1α polypeptide can be phosphorylated by p42/44 MAP kinases both under hypoxic conditions and in response to receptor-mediated ERK-activating stimuli [96–98]. The ERK-mediated phosphorylation was found to enhance the transcriptional activity of HIF-1 in various model systems, although the exact mechanism is still not clear [99, 100]. On one hand, it was proposed that phosphorylation of HIF-1α at positions 641 and 643 supports the transcriptional activity by attenuating its nuclear export

[101, 102]. On the other, it was also demonstrated that ERK activity fundamentally alters the predicted composition of HIF-1-containing nuclear complexes suggesting multiple effects of ERK activity on hypoxia signaling [103]. Independently of the mechanistic details, current data suggest that ERK-mediated upregulation of the HIF pathway differs from the mTOR-mediated effect and, primarily, acts on the transactivation function of HIFs, possibly, complementing the mTOR-mediated effects (**Figure 4**).

## 5. Conclusion

Extensive experimental work over the past three decades deciphered the molecular background of the cellular response to oxygen depletion, one of the fundamental physiologic processes. To date, these efforts depictured an intricate molecular network that bridges, apparently, every aspect of cellular physiology. Within this network, the PHD-HIF axis plays an integrative role of various signals that allows the hypoxic cell to shape dynamics of the adaptive response according to the actual endogenous metabolic state and surrounding microenvironment alike. Deeper understanding of these molecular machineries gives the opportunity to develop more efficient medical modalities for pathologies like chronic inflammation, ischemia or neoplasms.

## Acknowledgements

The Open Access fee of this chapter was funded by the Munchausen fund provided by the Department of Medical Chemistry, Molecular Biology, and Pathobiochemistry, Faculty of Medicine, Semmelweis University.

## Conflict of interest

The author declares no conflict of interest.

## Author details

Zsolt Fabian

Address all correspondence to: fabian.zsolt@med.semmelweis-univ.hu

Department of Medical Chemistry, Molecular Biology and Pathobiochemistry, Faculty of Medicine, Semmelweis University, Budapest, Hungary

# References

[1]  Conway Morris S. Darwin's dilemma: The realities of the Cambrian 'explosion'. Philosophical Transactions of the Royal Society of London. Series B, Biological Sciences. 2006;**361**(1470):1069-1083

[2]  Semenza GL. Hypoxia-inducible factor 1: Control of oxygen homeostasis in health and disease. Pediatric Research. 2001;**49**(5):614-617

[3]  Semenza GL et al. Hypoxia-inducible nuclear factors bind to an enhancer element located 3′ to the human erythropoietin gene. Proceedings of the National Academy of Sciences of the United States of America. 1991;**88**(13):5680-5684

[4]  Talks KL et al. The expression and distribution of the hypoxia-inducible factors HIF-1 alpha and HIF-2 alpha in normal human tissues, cancers, and tumor-associated macrophages. The American Journal of Pathology. 2000;**157**(2):411-421

[5]  Jain S et al. Expression of ARNT, ARNT2, HIF1 alpha, HIF2 alpha and Ah receptor mRNAs in the developing mouse. Mechanisms of Development. 1998;**73**(1):117-123

[6]  Jaakkola P et al. Targeting of HIF-alpha to the von Hippel-Lindau ubiquitylation complex by $O_2$-regulated prolyl hydroxylation. Science. 2001;**292**(5516):468-472

[7]  Bruick RK, McKnight SL. A conserved family of prolyl-4-hydroxylases that modify HIF. Science. 2001;**294**(5545):1337-1340

[8]  Willam C et al. The prolyl hydroxylase enzymes that act as oxygen sensors regulating destruction of hypoxia-inducible factor alpha. Advances in Enzyme Regulation. 2004;**44**:75-92

[9]  Hirsila M et al. Characterization of the human prolyl 4-hydroxylases that modify the hypoxia-inducible factor. The Journal of Biological Chemistry. 2003;**278**(33):30772-30780

[10]  Maxwell PH et al. The tumour suppressor protein VHL targets hypoxia-inducible factors for oxygen-dependent proteolysis. Nature. 1999;**399**(6733):271-275

[11]  Lando D et al. Asparagine hydroxylation of the HIF transactivation domain a hypoxic switch. Science. 2002;**295**(5556):858-861

[12]  Li SH et al. A novel mode of action of YC-1 in HIF inhibition: Stimulation of FIH-dependent p300 dissociation from HIF-1 {alpha}. Molecular Cancer Therapeutics. 2008;**7**(12):3729-3738

[13]  McNeill LA et al. Hypoxia-inducible factor asparaginyl hydroxylase (FIH-1) catalyses hydroxylation at the beta-carbon of asparagine-803. The Biochemical Journal. 2002; **367**(Pt 3):571-575

[14]  Semenza GL, Wang GL. A nuclear factor induced by hypoxia via de novo protein synthesis binds to the human erythropoietin gene enhancer at a site required for transcriptional activation. Molecular and Cellular Biology. 1992;**12**(12):5447-5454

[15] Wang GL, Semenza GL. Desferrioxamine induces erythropoietin gene expression and hypoxia-inducible factor 1 DNA-binding activity: Implications for models of hypoxia signal transduction. Blood. 1993;**82**(12):3610-3615

[16] Nemeth E et al. Hepcidin regulates cellular iron efflux by binding to ferroportin and inducing its internalization. Science. 2004;**306**(5704):2090-2093

[17] Ravasi G et al. Circulating factors are involved in hypoxia-induced hepcidin suppression. Blood Cells, Molecules and Diseases. 2014;**53**(4):204-210

[18] Mastrogiannaki M et al. Hepatic hypoxia-inducible factor-2 down-regulates hepcidin expression in mice through an erythropoietin-mediated increase in erythropoiesis. Haematologica. 2012;**97**(6):827-834

[19] Hintze KJ, McClung JP. Hepcidin: A critical regulator of iron metabolism during hypoxia. Advances in Hematology. 2011;**2011**:510304

[20] Ravasi G et al. Hepcidin regulation in a mouse model of acute hypoxia. European Journal of Haematology. 2018;**100**(6):636-643

[21] Leung PS et al. Increased duodenal iron uptake and transfer in a rat model of chronic hypoxia is accompanied by reduced hepcidin expression. Gut. 2005;**54**(10):1391-1395

[22] Shweiki D et al. Vascular endothelial growth factor induced by hypoxia may mediate hypoxia-initiated angiogenesis. Nature. 1992;**359**(6398):843-845

[23] Takeda N et al. Endothelial PAS domain protein 1 gene promotes angiogenesis through the transactivation of both vascular endothelial growth factor and its receptor, Flt-1. Circulation Research. 2004;**95**(2):146-153

[24] De Bock K et al. Role of PFKFB3-driven glycolysis in vessel sprouting. Cell. 2013;**154**(3): 651-663

[25] Xu Y et al. Endothelial PFKFB3 plays a critical role in angiogenesis. Arteriosclerosis, Thrombosis, and Vascular Biology. 2014;**34**(6):1231-1239

[26] McNeill LA et al. Hypoxia-inducible factor prolyl hydroxylase 2 has a high affinity for ferrous iron and 2-oxoglutarate. Molecular BioSystems. 2005;**1**(4):321-324

[27] Benita Y et al. An integrative genomics approach identifies Hypoxia Inducible Factor-1 (HIF-1)-target genes that form the core response to hypoxia. Nucleic Acids Research. 2009;**37**(14):4587-4602

[28] Linn TC, Pettit FH, Reed LJ. Alpha-keto acid dehydrogenase complexes. X. Regulation of the activity of the pyruvate dehydrogenase complex from beef kidney mitochondria by phosphorylation and dephosphorylation. Proceedings of the National Academy of Sciences of the United States of America. 1969;**62**(1):234-241

[29] Seifert F et al. Phosphorylation of serine 264 impedes active site accessibility in the E1 component of the human pyruvate dehydrogenase multienzyme complex. Biochemistry. 2007;**46**(21):6277-6287

[30] Pan Y et al. Multiple factors affecting cellular redox status and energy metabolism modulate hypoxia-inducible factor prolyl hydroxylase activity in vivo and in vitro. Molecular and Cellular Biology. 2007;**27**(3):912-925

[31] Selak MA et al. Succinate links TCA cycle dysfunction to oncogenesis by inhibiting HIF-alpha prolyl hydroxylase. Cancer Cell. 2005;**7**(1):77-85

[32] Tannahill GM et al. Succinate is an inflammatory signal that induces IL-1beta through HIF-1 alpha. Nature. 2013;**496**(7444):238-242

[33] Mracek T, Drahota Z, Houstek J. The function and the role of the mitochondrial glycerol-3-phosphate dehydrogenase in mammalian tissues. Biochimica et Biophysica Acta. 2013;**1827**(3):401-410

[34] Kelly TJ et al. A hypoxia-induced positive feedback loop promotes hypoxia-inducible factor 1 alpha stability through miR-210 suppression of glycerol-3-phosphate dehydrogenase 1-like. Molecular and Cellular Biology. 2011;**31**(13):2696-2706

[35] Hagen T et al. Redistribution of intracellular oxygen in hypoxia by nitric oxide: effect on HIF1 alpha. Science. 2003;**302**(5652):1975-1978

[36] Tanaka H et al. MicroRNA-183 upregulates HIF-1 alpha by targeting isocitrate dehydrogenase 2 (IDH2) in glioma cells. Journal of Neuro-Oncology. 2013;**111**(3):273-283

[37] Giannakakis A et al. miR-210 links hypoxia with cell cycle regulation and is deleted in human epithelial ovarian cancer. Cancer Biology and Therapy. 2008;**7**(2):255-264

[38] Marxsen JH et al. Hypoxia-inducible factor-1 (HIF-1) promotes its degradation by induction of HIF-alpha-prolyl-4-hydroxylases. The Biochemical Journal. 2004;**381**(Pt 3):761-767

[39] Stiehl DP et al. Increased prolyl 4-hydroxylase domain proteins compensate for decreased oxygen levels. Evidence for an autoregulatory oxygen-sensing system. Journal of Biological Chemistry. 2006;**281**(33):23482-23491

[40] Zhang Z et al. MicroRNA miR-210 modulates cellular response to hypoxia through the MYC antagonist MNT. Cell Cycle. 2009;**8**(17):2756-2768

[41] Kim JW et al. Hypoxia-inducible factor 1 and dysregulated c-Myc cooperatively induce vascular endothelial growth factor and metabolic switches hexokinase 2 and pyruvate dehydrogenase kinase 1. Molecular and Cellular Biology. 2007;**27**(21):7381-7393

[42] Doe MR et al. Myc posttranscriptionally induces HIF1 protein and target gene expression in normal and cancer cells. Cancer Research. 2012;**72**(4):949-957

[43] Zhang H et al. HIF-1 inhibits mitochondrial biogenesis and cellular respiration in VHL-deficient renal cell carcinoma by repression of C-MYC activity. Cancer Cell. 2007;**11**(5):407-420

[44] Koshiji M et al. HIF-1 alpha induces cell cycle arrest by functionally counteracting Myc. The EMBO Journal. 2004;**23**(9):1949-1956

[45] Yoshioka Y et al. Micromanaging iron homeostasis: Hypoxia-inducible micro-RNA-210 suppresses iron homeostasis-related proteins. The Journal of Biological Chemistry. 2012;**287**(41):34110-34119

[46] Wang H et al. Negative regulation of Hif1a expression and TH17 differentiation by the hypoxia-regulated microRNA miR-210. Nature Immunology. 2014;**15**(4):393-401

[47] Bruning U et al. MicroRNA-155 promotes resolution of hypoxia-inducible factor 1 alpha activity during prolonged hypoxia. Molecular and Cellular Biology. 2011;**31**(19):4087-4096

[48] Makino Y et al. Inhibitory PAS domain protein (IPAS) is a hypoxia-inducible splicing variant of the hypoxia-inducible factor-3 alpha locus. The Journal of Biological Chemistry. 2002;**277**(36):32405-32408

[49] Makino Y et al. Inhibitory PAS domain protein is a negative regulator of hypoxia-inducible gene expression. Nature. 2001;**414**(6863):550-554

[50] Makino Y et al. Transcriptional up-regulation of inhibitory PAS domain protein gene expression by hypoxia-inducible factor 1 (HIF-1): A negative feedback regulatory circuit in HIF-1-mediated signaling in hypoxic cells. The Journal of Biological Chemistry. 2007;**282**(19):14073-14082

[51] Reyes H, Reisz-Porszasz S, Hankinson O. Identification of the Ah receptor nuclear translocator protein (Arnt) as a component of the DNA binding form of the Ah receptor. Science. 1992;**256**(5060):1193-1195

[52] Poland A, Glover E, Kende AS. Stereospecific, high affinity binding of 2,3,7,8-tetrachlorodibenzo-p-dioxin by hepatic cytosol. Evidence that the binding species is receptor for induction of aryl hydrocarbon hydroxylase. The Journal of Biological Chemistry. 1976;**251**(16):4936-4946

[53] Perdew GH. Association of the Ah receptor with the 90-kDa heat shock protein. The Journal of Biological Chemistry. 1988;**263**(27):13802-13805

[54] Hoffman EC et al. Cloning of a factor required for activity of the Ah (dioxin) receptor. Science. 1991;**252**(5008):954-958

[55] Tomita S et al. Conditional disruption of the aryl hydrocarbon receptor nuclear translocator (Arnt) gene leads to loss of target gene induction by the aryl hydrocarbon receptor and hypoxia-inducible factor 1 alpha. Molecular Endocrinology. 2000;**14**(10):1674-1681

[56] Nukaya M et al. Aryl hydrocarbon receptor nuclear translocator in hepatocytes is required for aryl hydrocarbon receptor-mediated adaptive and toxic responses in liver. Toxicological Sciences. 2010;**118**(2):554-563

[57] Whitlock JP Jr et al. Cytochromes P450 5: Induction of cytochrome P4501A1: A model for analyzing mammalian gene transcription. The FASEB Journal. 1996;**10**(8):809-818

[58] Munzel PA et al. Aryl hydrocarbon receptor-inducible or constitutive expression of human UDP glucuronosyltransferase UGT1A6. Archives of Biochemistry and Biophysics. 1998;**350**(1):72-78

[59] Nie M, Blankenship AL, Giesy JP. Interactions between aryl hydrocarbon receptor (AhR) and hypoxia signaling pathways. Environmental Toxicology and Pharmacology. 2001;**10**(1-2):17-27

[60] Schults MA et al. Diminished carcinogen detoxification is a novel mechanism for hypoxia-inducible factor 1-mediated genetic instability. The Journal of Biological Chemistry. 2010;**285**(19):14558-14564

[61] Gradin K et al. Functional interference between hypoxia and dioxin signal transduction pathways: Competition for recruitment of the Arnt transcription factor. Molecular and Cellular Biology. 1996;**16**(10):5221-5231

[62] Chan WK et al. Cross-talk between the aryl hydrocarbon receptor and hypoxia inducible factor signaling pathways. Demonstration of competition and compensation. Journal of Biological Chemistry. 1999;**274**(17):12115-12123

[63] Mayerhofer M et al. BCR/ABL induces expression of vascular endothelial growth factor and its transcriptional activator, hypoxia inducible factor-1 alpha, through a pathway involving phosphoinositide 3-kinase and the mammalian target of rapamycin. Blood. 2002;**100**(10):3767-3775

[64] Treins C et al. Insulin stimulates hypoxia-inducible factor 1 through a phosphatidylinositol 3-kinase/target of rapamycin-dependent signaling pathway. The Journal of Biological Chemistry. 2002;**277**(31):27975-27981

[65] Humar R et al. Hypoxia enhances vascular cell proliferation and angiogenesis in vitro via rapamycin (mTOR)-dependent signaling. The FASEB Journal. 2002;**16**(8):771-780

[66] Land SC, Tee AR. Hypoxia-inducible factor 1 alpha is regulated by the mammalian target of rapamycin (mTOR) via an mTOR signaling motif. Journal of Biological Chemistry. 2007;**282**(28):20534-20543

[67] Pez F et al. The HIF-1-inducible lysyl oxidase activates HIF-1 via the Akt pathway in a positive regulation loop and synergizes with HIF-1 in promoting tumor cell growth. Cancer Research. 2011;**71**(5):1647-1657

[68] Heesom KJ, Denton RM. Dissociation of the eukaryotic initiation factor-4E/4E-BP1 complex involves phosphorylation of 4E-BP1 by an mTOR-associated kinase. FEBS Letters. 1999;**457**(3):489-493

[69] Magagnin MG et al. The mTOR target 4E-BP1 contributes to differential protein expression during normoxia and hypoxia through changes in mRNA translation efficiency. Proteomics. 2008;**8**(5):1019-1028

[70] Nayak BK et al. Stabilization of HIF-2 alpha through redox regulation of mTORC2 activation and initiation of mRNA translation. Oncogene. 2013;**32**(26):3147-3155

[71] Moon EJ et al. NADPH oxidase-mediated reactive oxygen species production activates hypoxia-inducible factor-1 (HIF-1) via the ERK pathway after hyperthermia treatment.

Proceedings of the National Academy of Sciences of the United States of America. 2010;**107**(47):20477-20482

[72] Bonello S et al. Reactive oxygen species activate the HIF-1 alpha promoter via a functional NFkappaB site. Arteriosclerosis, Thrombosis, and Vascular Biology. 2007;**27**(4):755-761

[73] Simon MC. Mitochondrial reactive oxygen species are required for hypoxic HIF alpha stabilization. Advances in Experimental Medicine and Biology. 2006;**588**:165-170

[74] Demidenko ZN, Blagosklonny MV. The purpose of the HIF-1/PHD feedback loop: To limit mTOR-induced HIF-1 alpha. Cell Cycle. 2011;**10**(10):1557-1562

[75] Cassavaugh JM et al. Negative regulation of HIF-1 alpha by an FBW7-mediated degradation pathway during hypoxia. Journal of Cellular Biochemistry. 2011;**112**(12):3882-3890

[76] Welcker M, Clurman BE. FBW7 ubiquitin ligase: A tumour suppressor at the crossroads of cell division, growth and differentiation. Nature Reviews. Cancer. 2008;**8**(2):83-93

[77] Sakamoto T et al. Hypoxia-inducible factor 1 regulation through cross talk between mTOR and MT1-MMP. Molecular and Cellular Biology. 2014;**34**(1):30-42

[78] Lando D et al. FIH-1 is an asparaginyl hydroxylase enzyme that regulates the transcriptional activity of hypoxia-inducible factor. Genes and Development. 2002;**16**(12):1466-1471

[79] Petrella BL, Lohi J, Brinckerhoff CE. Identification of membrane type-1 matrix metalloproteinase as a target of hypoxia-inducible factor-2 alpha in von Hippel-Lindau renal cell carcinoma. Oncogene. 2005;**24**(6):1043-1052

[80] Brown EJ et al. Control of p70 s6 kinase by kinase activity of FRAP in vivo. Nature. 1995;**377**(6548):441-446

[81] Brugarolas J et al. Regulation of mTOR function in response to hypoxia by REDD1 and the TSC1/TSC2 tumor suppressor complex. Genes and Development. 2004;**18**(23):2893-2904

[82] Inoki K et al. Rheb GTPase is a direct target of TSC2 GAP activity and regulates mTOR signaling. Genes and Development. 2003;**17**(15):1829-1834

[83] Li Y et al. Bnip3 mediates the hypoxia-induced inhibition on mammalian target of rapamycin by interacting with Rheb. The Journal of Biological Chemistry. 2007;**282**(49):35803-35813

[84] Wan G et al. Hypoxia-induced MIR155 is a potent autophagy inducer by targeting multiple players in the MTOR pathway. Autophagy. 2014;**10**(1):70-79

[85] Rius J et al. NF-kappaB links innate immunity to the hypoxic response through transcriptional regulation of HIF-1 alpha. Nature. 2008;**453**(7196):807-811

[86] Minet E et al. HIF1A gene transcription is dependent on a core promoter sequence encompassing activating and inhibiting sequences located upstream from the transcription initiation site and cis elements located within the 5' UTR. Biochemical and Biophysical Research Communications. 1999;**261**(2):534-540

[87]  Fitzpatrick SF et al. An intact canonical NF-kappaB pathway is required for inflammatory gene expression in response to hypoxia. Journal of Immunology. 2011;**186**(2):1091-1096

[88]  Cummins EP et al. Prolyl hydroxylase-1 negatively regulates IkappaB kinase-beta, giving insight into hypoxia-induced NFkappaB activity. Proceedings of the National Academy of Sciences of the United States of America. 2006;**103**(48):18154-18159

[89]  Scholz CC et al. Regulation of IL-1beta-induced NF-kappaB by hydroxylases links key hypoxic and inflammatory signaling pathways. Proceedings of the National Academy of Sciences of the United States of America. 2013;**110**(46):18490-18495

[90]  van Uden P et al. Evolutionary conserved regulation of HIF-1beta by NF-kappaB. PLoS Genetics. 2011;**7**(1):e1001285

[91]  Fitzpatrick SF et al. Prolyl hydroxylase-1 regulates hepatocyte apoptosis in an NF-kappaB-dependent manner. Biochemical and Biophysical Research Communications. 2016; **474**(3):579-586

[92]  Oliver KM et al. Hypoxia activates NF-kappaB-dependent gene expression through the canonical signaling pathway. Antioxidants and Redox Signaling. 2009;**11**(9):2057-2064

[93]  Schmedtje JFJ et al. Hypoxia induces cyclooxygenase-2 via the NF-kappaB p65 transcription factor in human vascular endothelial cells. The Journal of Biological Chemistry. 1997;**272**(1):601-608

[94]  Figueroa YG et al. NF-kappaB plays a key role in hypoxia-inducible factor-1-regulated erythropoietin gene expression. Experimental Hematology. 2002;**30**(12):1419-1427

[95]  Bandarra D et al. HIF-1 alpha restricts NF-kappaB-dependent gene expression to control innate immunity signals. Disease Models and Mechanisms. 2015;**8**(2):169-181

[96]  Richard DE et al. p42/p44 mitogen-activated protein kinases phosphorylate hypoxia-inducible factor 1 alpha (HIF-1 alpha) and enhance the transcriptional activity of HIF-1. The Journal of Biological Chemistry. 1999;**274**(46):32631-32637

[97]  Minet E et al. ERK activation upon hypoxia: Involvement in HIF-1 activation. FEBS Letters. 2000;**468**(1):53-58

[98]  Sodhi A et al. The Kaposi's sarcoma-associated herpes virus G protein-coupled receptor up-regulates vascular endothelial growth factor expression and secretion through mitogen-activated protein kinase and p38 pathways acting on hypoxia-inducible factor 1 alpha. Cancer Research. 2000;**60**(17):4873-4880

[99]  Shi YH et al. In vitro study of HIF-1 activation and VEGF release by bFGF in the T47D breast cancer cell line under normoxic conditions: Involvement of PI-3K/Akt and MEK1/ERK pathways. The Journal of Pathology. 2005;**205**(4):530-536

[100]  Dimova EY, Kietzmann T. The MAPK pathway and HIF-1 are involved in the induction of the human PAI-1 gene expression by insulin in the human hepatoma cell line HepG2. Annals of the New York Academy of Sciences. 2006;**1090**:355-367

[101] Mylonis I et al. Identification of MAPK phosphorylation sites and their role in the localization and activity of hypoxia-inducible factor-1 alpha. The Journal of Biological Chemistry. 2006;**281**(44):33095-33106

[102] Mylonis I et al. Atypical CRM1-dependent nuclear export signal mediates regulation of hypoxia-inducible factor-1 alpha by MAPK. The Journal of Biological Chemistry. 2008;**283**(41):27620-27627

[103] Fabian Z et al. Basic fibroblast growth factor modifies the hypoxic response of human bone marrow stromal cells by ERK-mediated enhancement of HIF-1 alpha activity. Stem Cell Research. 2014;**12**(3):646-658

# Chapter 5

# Vitamin K2: A Vitamin that Works like a Hormone, Impinging on Gene Expression

Jan Oxholm Gordeladze

Additional information is available at the end of the chapter

http://dx.doi.org/10.5772/intechopen.80388

### Abstract

Vitamin K2 binds to the intranuclear receptor SXR and results in the activation of a plethora of genes, both directly and indirectly. Among these genes are important biological markers of cellular characteristics or features (also known as cell phenotypes), as well as a set of molecules known to be involved in both hormone-induced, G-protein-mediated cell signalling, either directly or indirectly activating so-called sirtuins and/or histone deacetylaces (HDACs), known as determinants of cell types and their specific functions in a given tissue. Hence, vitamin K2 may be closely involved in or serving as a traditional molecular 'link' between hormonal receptors and intracellular signalling pathways. It has been stated that a true hormone is a product of living cells, which circulates in body fluids (such as blood) and elicits a specific and often stimulatory effect on the activity of cells situated remotely from its point of origin. A large bulk of evidence published over the past 10 years establishes vitamin K2 in this category of substances. Hence, vitamin K2 should be considered and consequently classified as a hormone.

**Keywords:** vitamin K2, SXR, G-proteins, vitamin A-D-K2 cascade, bioinformatics, in vitro model systems

## 1. Introduction

Vitamin K has since more than 25 years been known to serve as a powerful nutrient factor in the preservation of homeostatic bone turnover, along with blood clotting biochemistry. In addition, vitamin K has been used as a therapeutic remedy in the clinic to treat and prevent bone brittleness (osteoporosis) in Japan and many other countries around the world. Moreover, beyond its enzymatic character as a cofactor for the vitamin K-dependent GGCX (gamma-glutamyl carboxylase), Professor Inoue and his co-workers have shown that the K2 variant is, in fact, a transcriptional modulator of osseous marker genes, also serving as an extracellular matrix-related molecule, via the stimulation of the 'steroid and xenobiotic receptor' SXR [1]. A microarray-based revelation of the present action of vitamin K2 in bone-derived osteoblastic cell genes corroborated the notion that the K2 variant menaquinone-4 (MK-4), in fact, is a hormone. Among the significantly upregulated genes, both growth differentiation factor 15 (GDF15) and stanniocalcin 2 (STC2) were concluded to serve as novel target genes, which both circumvented the traditional GGCX- and SXR-mediated pathways found in osteoblast-like cells.

## 2. Vitamin K2: vitamin and hormone

The induction of both GDF15 and STC2 genes is construed as specific to MK-4, since it was shown not to be brought about by another vitamin K(2) isoform MK-7, vitamin K(1) or the MK-4 side chain containing the geranylgeraniol group. A survey into the signalling pathways in question indicated that MK-4 sustained phosphorylation of protein kinase A (PKA) and that MK-4 mediated upregulation of genes, such as GDF15 and STC2, was diminished by the exposure to a PKA inhibitor or by siRNA constructs against PKA. These observations were in line with the concept that vitamin K(2) was capable of modulating its own target gene expression in bone-derived (osteoblastic) cell entities through a PKA-driven pathway, which in essence was different from the traditional vitamin K-dependent signalling pathways [2].

Vitamin K2 has been included as a member in a group of molecules, constituting the 'requirement' for blood to coagulate; however, it has been demonstrated to function or serve as a key in the homeostasis of osseous tissue, thus showing effectiveness as one therapeutic agent, among others, in the curative treatment portfolio of bone ailments and diseases, i.e. like bone loss or brittleness (osteomalacia and osteoporosis). Furthermore, it has, since several decades, been acknowledged that vitamin K2 mediates transcriptional modulation of marker genes in osteoblastic cells, as well as reinforcing bone formation via the nuclear steroid and xenobiotic receptor, SXR. In this context, Dr. Satoshi Inoue and his research team identified several genes, which were upregulated by vitamin K2 (and a prototypical SXR like ligand, rifampicin) in osteoblastic cells, through microarray analysis and PCR. A plethora of genes was upregulated, among which collagen synthesis and accumulation in osteoblast-like MG63 cells were enhanced several times over by vitamin K2 treatment. Therefore, the results of Dr. Inoue and his research group more than suggested a novel function for vitamin K2 in the formation of osseous tissues, i.e. that K2 was a true transcriptional regulator of extracellular matrix-related genes, being involved in the assembly of collagen. At present, we know that vitamin K2 (or menaquinone-7 = MK-7, among other vitamin K2 metabolites) works through this nuclear receptor and consequently should be classified as a hormone and not solely be construed as 'a vitamin'.

Arterial stiffness is always associated with an enhanced cardiovascular risk, morbidity and mortality [3]. The present article reviews the main vitamins being involved in arterial stiffness and enabling destiffening; their mechanism of action, providing a brief description of the latest studies in the area; and their implications for primary cardiovascular prevention, clinical practice and therapy. Despite inconsistent proof for 'softening' brought about by vitamin supplementation in a plethora of clinical trials, promising results were observed in selected populations. The chief mechanisms pertain to anti-atherogenic potential, substantial augmentation of endothelial functionality (pertaining to vitamins A, C, D and E, respectively) and general metabolic profiling (pertaining to vitamins A, B12, C, D and K, respectively), suppression of the renin-angiotensin-aldosterone (R-R-A) system (vitamin D), anti-inflammatory (vitamins A-D-E-K) and antioxidant effects (vitamins A-C-E), diminished homocysteine levels (vitamin B12) and a reversal of the calcification of arteries (vitamin K). Vitamins A, B12, C, D and E, as well as vitamin K status, are important in evaluating the risk of cardiovascular events, and, finally, supplementation with vitamins may serve as an efficient, individually based and less costly 'destiffening' therapeutic mode.

Vitamin K is renowned for being an important (vital) nutrient, sustaining both bone homeostasis and blood coagulation. Therefore, it is both clinically and ubiquitously used as a therapeutic agent or treatment for osteoporosis in Japan and western countries. Besides its powerful enzymatic, cofactor role of the vitamin K-dependent g-glutamyl carboxylase (GGCX), it has since long been known that vitamin K2 may serve as a transcriptional regulator of marker osteoblastic genes, as well as in matrix-related, extracellular genes. In this context the activation of the so-called steroid and xenobiotic receptor (known as SXR) is mandatory.

Hence, genes known to be upregulated by vitamin K2 isoforms like menaquinone-4 (MK-4) were applied using oligonucleotide-based microarray analyses. Among the MK-4 upregulated gene species, the growth differentiation factor 15 (GDF15) and stanniocalcin 2 (STC2) were discovered as new MK-4 target genes, being independent of the GGCX and SXR pathways in osteoblastic cells. The observed induction of GDF15 and STC2 was construed as specific to MK-4, since it was not seen with exposure to MK-7 or vitamin K1. Surprisingly, a scrutiny of the main signalling pathways showed that MK-4 stimulated PKA (protein kinase A) phosphorylation. Furthermore, the MK-4-dependent induction of both GDF15 and STC2 genes was obliterated subsequent to the treatment with PKA inhibitors or siRNAs against PKA. Therefore, it was postulated that vitamin K2 (MK-7) could modulate target gene expression in osteoblastic cells via PKA-dependent mechanisms, which were distinct from any previously known vitamin K-mediated signalling pathway.

The paper found that vitamin K2 is recognised, along with calcium, vitamin D and magnesium, as essential in supporting strong bones and healthy arteries. In the paper, *Nutritional strategies for skeletal and cardiovascular health: hard bones, soft arteries, rather than vice versa*, the authors cite a US Surgeon General's Report that states that one in two Americans over 50 is expected to have or to be at risk of developing osteoporosis, which causes 8.9 million fractures annually, with an estimated cumulative cost of incident fractures predicted at $474 billion during the next 20 years in the USA.

Furthermore, a study conducted by the Mayo Clinic [4, 5] reported that 'compared with 30 years ago, forearm fractures have risen more than 32% in boys and 56% in girls'. Meanwhile, strong epidemiological associations exist between decreased bone mineral density (BMD) and increased risk of cardiovascular (CV) disease. For example, individuals with osteoporosis have a higher risk of coronary artery disease and vice versa. This problem will be magnified, according to the paper, if the therapies for osteoporosis (calcium supplements) independently increase risk of myocardial infarction. To that end, the authors conducted a comprehensive and systematic review of the scientific literature to determine the optimal dietary strategies and nutritional supplements for long-term skeletal health and cardiovascular health. They summarised what is helpful for building strong bones while maintaining soft and supple arteries:

- Obtain calcium from dietary sources (the best choice is a calcium hydroxyapatite) with the adequate animal protein, fruit and vegetable intake.

- Concomitantly increase potassium consumption, while reducing sodium intake should be taken into account.

- Maintain vitamin D levels in the normal range.

- Increase the intake of foods rich in vitamins K1 and K2.

The study notes: 'A meta-analysis concluded that while supplementation with phytonadione (vitamin K1) improved bone health, vitamin K2 was even more effective in this regard. This large and statistically rigorous meta-analysis concluded that high vitamin K2 levels were associated with reduced vertebral fractures by approximately 60%, hip fractures by 77% and all non-vertebral fractures by approximately 81%. Supplementation with vitamin K2 as MK-7 increased bone strength in postmenopausal women in 3-year clinical study'.

'Additionally, increased vitamin K2 intake has been associated with decreased arterial calcium deposition and the ability to reverse vascular calcification in animal models. Moreover clinical trial proved that vitamin K2 supplementation increases elasticity of the arteries (in 3 years)', the paper stated. The authors recommend increasing the intake of foods rich in vitamins K1 and K2 to secure skeletal and cardiovascular health. 'The positive health potential of vitamin K2 is more effective than for vitamin K1, the paper reads. Yet, Dr. Hogne Vik, Chief Medical Officer with NattoPharma, world leader in vitamin K2 R&D, exclusive global supplier of MenaQ7 vitamin K2 as MK-7, and sponsor of the 3-year studies cited in the paper, explains that it is not possible to get sufficient amounts of vitamin K2 through a European or US diet'.

The only food that contains enough vitamin K2 is the Japanese dish Natto. 'This means that if you want to get enough vitamin K2 into your body, then you have to take dietary supplements or functional foods containing vitamin K2', he said. 'We are gratified, but not surprised, that our 3-year clinical studies were cited in this paper', Dr. Vik continues. 'NattoPharma has driven the clinical research that has demonstrated vitamin K2's benefits for human health, and our breakthrough studies provided the first intervention data confirming the associations that observational studies made previously: that vitamin K2 as MK-7 is available beyond the liver to support bone and cardiovascular health. And it does this by activating proteins that help the body to properly utilise calcium – there by simultaneously supporting both skeletal and cardiovascular health'.

How are we going to interpret the above-described information about the mechanism of action of vitamin K2? Well, according to the definition given in dictionaries and scientific papers, vitamin K2 (as in the form or MK-4 or MK-7) fits this definition and should be classified as one.

## 3. What features characterise a true hormone?

To make the picture clearer, let us start all over again. According to **Figure 1**, the impact of a hormone is one signal among differentiation signals (also called epigenator factors), which may be temperature variations, oxygen tension, mechano-stimulation or humoral factor/hormones. The 'hormone' eventually activates a transcription factor or microRNA synthesis, which may impinge upon DNA in terms of a certain or given spectrum of mRNAs appearing in the cytosol of the cell. Of major interest here are eventually two classes of molecules utmost important for the acquisition of the final cell phenotype—the histones and the sirtuins (see **Figure 5**).

The response elicited by a single hormone (epigenator) may look like the network of interacting factors (mostly transcription factors), such as the network representing the closely cooperating network (mostly represented by transcription factor), as seen in **Figure 2a** and **b**.

The reader is recommended to look up the remainder of the genes shown above and will be amazed as to the plethora of biological effects being modulated (directly and/or indirectly) by the nuclear receptor NR1/NR2 = SXR, to which vitamin K2 binds, exerting its multitude of biological effects. The crucial question is then: what may be the immediate effect of vitamin K2 (e.g. MK-7) on cell phenotype, for example, on bone chips harbouring live osteoblastic cells? In **Figure 3**, bovine chips from young calves were incubated for 14 days in growth medium, which were analysed for osteocalcin, IL-10, TGFβ, OPG and RANKL (osteoblast and osteoclast markers), respectively. For detailed summary of results, see **Figure 3**.

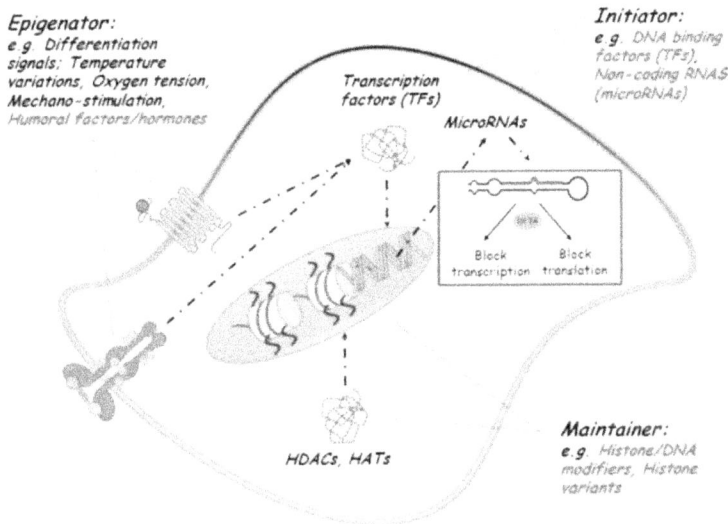

**Figure 1.** The epigenator-initiator-maintainer model of hormonal impact on the phenotype of a given cell. The hormone (binding to a given receptor—here represented by two different transmembrane proteins) will eventually elicit a response determined by the joint effect of transcription factors and microRNAs determining the end-point effect of the epigenator, such as cell differentiation and/or efflux of secretory products.

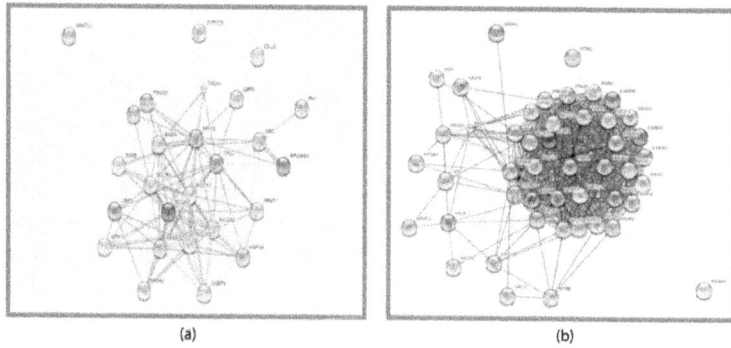

(a)                                                        (b)

**Figure 2.** (a) and (b) GeneCards-based emulation of molecules interacting with NR1/NR2 (which is identical to SXR, identified/described by the Japanese researcher, Professor Satoshi Inoue). Interestingly, NR1/NR2 is integrated in a large network of interacting molecules (genes), representing various classes of transcription factors (e.g. PPARG, RORC, RARA, RXR) and not to forget the thyroid receptors A and B!

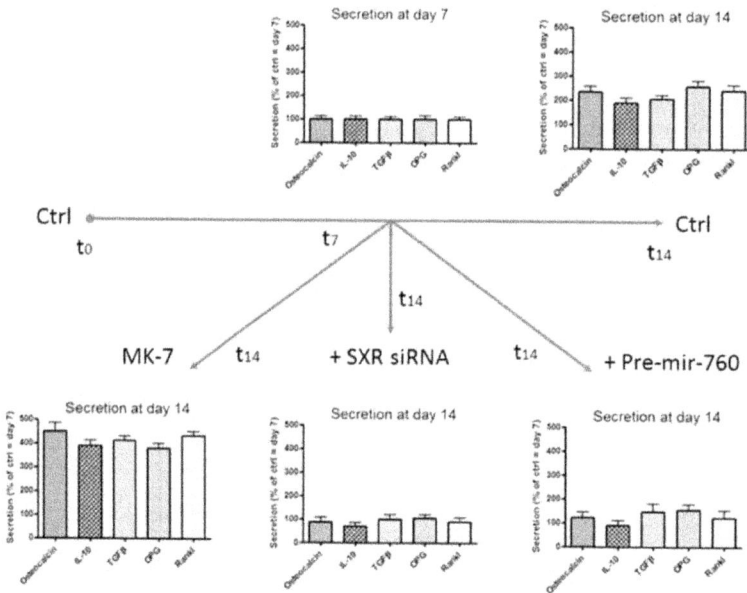

**Figure 3.** Bovine chips from young calves incubated as stated above. The chips were (a) incubated for 7 or 21 days and analysed for osteocalcin, IL-10, TGFβ, OPG and RANKL (all parameters' characteristics for the osteoblast phenotype); (b) incubated for 7 days and thereafter for 14 days in the presence of either vitamin K2 = MK-7; siRNA against the vitamin K2 receptor SXR or with pre-mir-760. The final measurements of secretory products were as indicated above (osteocalcin, IL-10, TGFβ, OPG and RANKL).

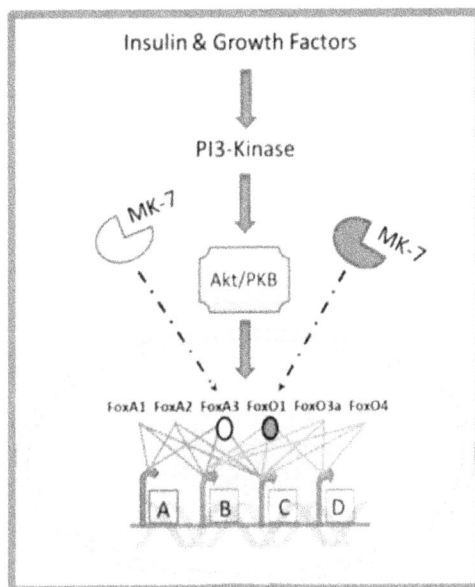

**Figure 4.** The impact of insulin and growth factors (GFs) on members of the FoxA and FoxO families of insulin and growth factors.

The interpretation of the experiment with 'live' bovine bone chips is the following: since MK-7 (a vitamin K2 analogue) is stimulated, while SXR siRNA and pre-mir-760 markedly reduced the secretory profile of the bovine osteoblast, it was concluded that the effect of vitamin K2 (MK-7) on the osteoblast profile of secretory molecules was stimulatory and that the effect of vitamin K2 was mediated solely through the action of the nuclear receptor SXR.

Furthermore, it was shown that one well-known and well-represented vitamin K2 'analogue' (MK-7) was able to activate (upregulate) some of the members of the FoxA and FoxO family of transcription factors. In our hands, vitamin K2 analogue MK-7 (and to a lesser extent MK-4) was able to substitute for insulin, as well as some growth factors (like growth hormone and a few interleukins (not shown)) (**Figure 4**).

**Figure 5** shows a strong relationship (network between a human osteoblast microRNA profile, respectively, and a well-known plethora of marker genes) which has previously been listed according to their sequential appearance as the osteoblast develops from a stem cell into a 'full-blown', mature mineralizing osteoblast. By adding a plethora of transcription factors, sirtuins (SIRTs), histone deacetylases (HDACs) and transcription factors (with special references to the osteoblastic phenotype), a distinct pattern of interaction appeared. By adding vitamin K2, i.e. MK-7 and/or SXR, to the system, a similar pattern appeared (not shown[1]). This was also the case for other tissues as well, e.g. lung, heart, brain, muscle, white/beige fat tissues and many others (not mentioned here[2]).

---

[1]Details not included—patents pending.
[2]Details not included—patents pending.

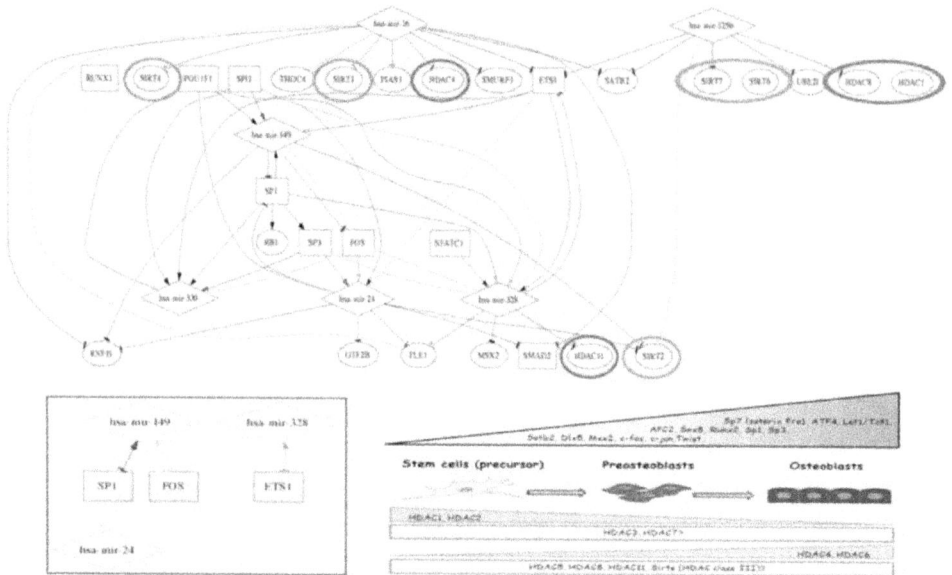

**Figure 5.** Network (based on the computerprogram Mir@nt@n) postulating the interaction between key microRNAs, transcription factors, sirtuins, and histone deacetylases (HDACs) in the developing osteoblast from stem cells via preosteoblasts. The compilation of genes was based on various searches on PubMed for genes appearing during the course of preosteoblast to mature (mineralizing) osteoblasts from mammalian species, including humans.

## Author details

Jan Oxholm Gordeladze

Address all correspondence to: j.o.gordeladze@medisin.uio.no

Department of Molecular Medicine, Section for Biochemistry, Institute of Basic Medical Science, Blindern, Norway

## References

[1] Sultana H, Watanabe K, Md MR, Takashima R, Ai O, Komai M, et al. Effects of Vitamin K2 on the Expression of Genes Involved in Bile Acid Synthesis and Glucose Homeostasis in Mice with Humanized PXR. Sendai, Japan: Laboratory of Nutrition, Graduate School of Agricultural Science, Tohoku University. 2018

[2] Ichikawa T, Horie-Inoue K, Ikeda K, Blumberg B, Inoue S. Vitamin K2 induces phosphorylation of protein kinase A and expression of novel target genes in osteoblastic cells. Journal of Molecular Endocrinology. 2007;**39**(4):239-247

[3]  Mosos I. Review article—Crosstalk between vitamins A, B12, D, K, C, and E status and arterial stiffness. Disease Markers. 2017;**2017**:14. Article ID: 8784971.  DOI: 10.1155/2017/8784971

[4]  https://www.bodyscience.com.au/blog/dr-hogne-vik-vitamin-k2-interview/

[5]  Gordeladze JO. General Reading on Vitamin K2: Vitamin K2—Vital for Health and Well-Being. IntechOpen. 2017. DOI: 10.5772/61430

www.ingramcontent.com/pod-product-compliance
Lightning Source LLC
Chambersburg PA
CBHW081242190326
41458CB00016B/5879